THEORY
OF FUNCTIONS

By

DR. KONRAD KNOPP

Professor of Mathematics at the University of Tübingen

Translated by

FREDERICK BAGEMIHL, M.A.

Instructor in Mathematics at the University of Rochester

PART ONE
ELEMENTS OF THE GENERAL THEORY OF
ANALYTIC FUNCTIONS

NEW YORK
DOVER PUBLICATIONS

PREFACE

This little ∫book follows rather closely the fifth edition of Dr. Knopp's *Funktionentheorie*. Several changes have been made in order to conform to common English terminology and notation, and to render certain passages more precise or rigorous than they are in the German volume. The proofs of Lemmas 1 and 2 in § 4 were found to be incorrect, and proofs remedying this defect were substituted. Typographical errors have been corrected, the bibliography has undergone some minor changes, and a few helpful references have been added to the text.

The Translator

CONTENTS

Section I

Fundamental Concepts

Section II

Integral Theorems

v

SECTION IV

Singularities

FUNDAMENTAL CONCEPTS

CHAPTER 1

NUMBERS AND POINTS

§1. Prerequisites

We presume that the reader is familiar with the theory of real numbers, with the foundations of real analysis (infinitesimal analysis, i.e., differential and integral calculus) which is built upon that theory, and with the elements of analytic geometry. The extent to which this is necessary in order to understand the subsequent presentation is amplified in the opening paragraphs of the *Elem.*[1] We suppose further that the reader is also familiar with the remaining contents of the *Elem.* Thus, we take for granted that he is acquainted with the ordinary complex numbers and that he is able to operate with them. It is assumed that he knows how the totality of these numbers[2] can be put into one-to-one correspondence with the points or vectors of a plane or with the points of a sphere, and how thereby every analytical consideration can be interpreted geometrically and every geometrical consideration followed analytically (*Elem.*, sec. I). We likewise take for granted that he is already acquainted, in the main, with infinite sequences and infinite series with complex terms, and with the concept of a function of a complex argument. We presume that he is familiar

---

[1] By "*Elem.*" we refer to the little volume *Elemente der Funktionentheorie*, Sammlung Göschen No. 1109. Berlin and Leipzig, 1937. Much of the material in the *Elem.* is to be found in G. H. Hardy, *A Course of Pure Mathematics*, 7th ed., New York, 1941, or in R. Courant, *Differential and Integral Calculus*, New York, 1938 (see especially Vol. I, Chapter I, and Appendix I, §§1 and 2; Chapter VIII; Vol. II, §1 of the Appendix to Chapter II, Chapter VIII, §§1 and 2).

[2] When we speak of "numbers" in the following, we mean the ordinary complex numbers unless it is expressly stated to the contrary.

with the application of the concept of limit to both, and consequently also with the concepts of continuity and differentiability of functions of a complex variable (*Elem.*, secs. III and IV). Finally, we suppose that he knows the most important properties of the so-called elementary functions (*Elem.*, secs. II and V).

Those topics of the *Elem.* which are most important for the present purposes will be reviewed and, in some cases, supplemented in this and the next chapter. The reader will thus be able to check for himself to what extent he possesses these prerequisites. At the same time, he will gain a firm basis for the subsequent development of the general theory of analytic functions.

§2. Plane and Sphere of Complex Numbers

The set of complex numbers can be put into one-to-one correspondence with the points of a plane oriented by a rectangular coordinate system. The plane is then called "the (Gaussian or complex) number plane" or, more briefly, "the z-plane." Every complex number $z = x + iy$ corresponds to that point whose abscissa is the real part $x = \Re(z)$ and whose ordinate is the imaginary part $y = \Im(z) = \Re(-iz)$.[1] As a consequence of this convention, precisely one point of the z-plane corresponds to every complex number z; and, conversely, precisely one complex number corresponds to every point of this plane. "Point" and "number" can therefore be used as equivalent expressions without fear of misinterpretations, so that we may use such expressions as "the point $i\sqrt{3}$," or "the distance between two numbers," or "the triangle with the vertices z_1, z_2, z_3," etc.

If r and φ are the polar coordinates of the point z,

---

[1] Small Roman or Greek (occasionally also German) characters always denote complex numbers if the contrary does not follow clearly from the context. Nevertheless, x, y, and later more frequently u, v, and ξ, η will be reserved for the real and imaginary part, respectively, and consequently, for real numbers. At times, iy (not y alone) is also used for the imaginary part of z. The context always excludes ambiguities.

then r is called the *absolute value* or *modulus* and φ the *amplitude*[1] of z. In symbols: $|z| = r$, am $z = \varphi$.

It is useful to call special attention to the following simple facts which follow from this equivalence of point and number.

a) The distance of a point z from the origin is $|z|$. The distance between two points z_1 and z_2 is $|z_1 - z_2| = |z_2 - z_1|$. The number $z_2 - z_1$ is represented by the vector extending from the point z_1 to the point z_2. The relations

$$|z_1 \pm z_2| \leqq |z_1| + |z_2| \text{ and } |z_1 \pm z_2| \geqq \big||z_1| - |z_2|\big|$$

hold for arbitrary z_1 and z_2.

b) The circumference of the circle of unit radius about the origin as center (the so-called **unit circle**) is characterized by $|z| = 1$; i.e., all numbers z for which $|z| = 1$ are points of this circumference, and conversely.

c) The interior of the circle of radius r about z_0 as center, exclusive of its circumference (its *boundary*), is characterized by $|z - z_0| < r$.

d) The interior of the circle of radius 3 about $-4i$ as center, inclusive of its boundary, is characterized by $|z + 4i| \leqq 3$.

e) That part of the z-plane which lies outside the circle of radius R about z_1 as center is given by $|z - z_1| > R$.

f) The *"right"* half-plane, i.e., that part of the z-plane which lies to the right of the imaginary axis in the usual orientation of the coordinate axes, exclusive of its boundary, is characterized by $\Re(z) > 0$. Likewise, the "upper" half-plane, inclusive of its boundary, is given by $\Im(z) \geqq 0$.

g) The interior of the circular ring formed by the circles of radii r and R about z_0 as center, exclusive of both boundaries, is represented by $0 < r < |z - z_0| < R$.

---

[1] The term "argument" (arg $z = \varphi$) is also in use.

h) A circle with radius ϵ about ζ as center, briefly called "a neighborhood" or more precisely "an ϵ-neighborhood" of the point ζ, consists of the points $\zeta + z'$ with fixed ζ and arbitrary z' subject only to the restriction $|z'| < \epsilon$ (compare **c**)). For, setting $\zeta + z' = z$, this means precisely that

$$|z'| = |z - \zeta| < \epsilon.$$

The plane of complex numbers is closed by introducing an improper point, the point[1] $z = \infty$ (see *Elem.*, §§14, 15, and 17). Therefore the exterior of a circle (cf. **e**)) is also called "a neighborhood of the point ∞." For the present, however, a letter will never denote the point ∞ if the contrary is not expressly stated.

By means of the so-called "stereographic projection" (see *Elem.*, ch. 3), the points of the complex plane are mapped one-to-one onto the points of a sphere called the *Riemann sphere*, the *sphere of complex numbers*, or briefly the *z-sphere*.

The customary way of doing this is as follows. A sphere of unit diameter is placed upon the z-plane in such a manner that the point of contact (south pole) lies at the origin. By means of rays emanating from the north pole, every point of the z-plane can be made to correspond, in a one-to-one fashion, to a point of the sphere. This point is again called briefly the point z of the sphere. The north pole of the sphere is then the representative (here entirely proper) of "the point ∞" of the z-plane. The complex plane which is closed by the point ∞ is said to have the same connectivity (the same topological structure) as the full sphere.

The equator of the sphere corresponds to the unit circle of the plane; the anterior (posterior) hemisphere, to the lower (upper) half-plane. The semi-meridians

---

[1] Note the difference between this and the following: (1) the set of real numbers (the real axis) which leads to the introduction of two improper values. $+ \infty$ and $- \infty$, and (2) the "projective plane" in which an infinite number of improper points are introduced. Structurally (topologically) the complex plane is intrinsically different from the projective plane.

correspond to the half-rays emanating from O; the parallels of latitude, to the circles about O as center.

An (ordinary) reflection about the equatorial plane is the same as an inversion with respect to the unit circle. The southern (northern) hemisphere maps into the interior (exterior) of the unit circle; a spherical cap about the north pole, into a neighborhood of the point ∞; etc.

Exercises. 1. Which curves in the plane are characterized by the following relations:

$\alpha)$ $\left|\dfrac{z-1}{z+1}\right| = 1$, $\beta)$ $\left|\dfrac{z-1}{z+1}\right| = 2$, $\gamma)$ $\left|\dfrac{z}{z+1}\right| = \alpha \,(> 0)$,

$\delta)$ $\Re(z^2) = 4$, $\epsilon)$ $\Im(z^2) = 4$, $\zeta)$ $|z^2 - 1| = \alpha \,(> 0)$?

Which parts of the plane are characterized by the same relations if the equality sign in them is replaced by $<, >, \leqq, \geqq$?

2. What relative positions in the plane or on the sphere do the following points have:

$a)$ z and $-z$; $b)$ z and $\bar{z}$;[1] $c)$ z and $-\bar{z}$;

$d)$ z and $\dfrac{1}{z}$; $e)$ z and $\dfrac{1}{\bar{z}}$; $f)$ z and $-\dfrac{1}{\bar{z}}$?

§3. Point Sets and Sets of Numbers

If a finite or an infinite number of complex numbers are selected according to any rule, these constitute a *set of numbers* and the corresponding points constitute a *point set*. "Point set" and "set of numbers" are considered as fully equivalent expressions. Such a set of numbers, $\mathfrak{M}$, is regarded as given or defined if its definition (the rule for selecting) enables one to decide whether a given number belongs to the set or not (and only the one or the other alternative is possible). Since the point set $\mathfrak{M}$ representing this set of numbers lies in the complex plane, one also speaks of "plane sets." The numbers (points) of the set are called its elements.

---

[1] The complex number which is the conjugate of z is denoted by $\bar{z}$. (If $z = x + iy$, $\bar{z} = x - iy$).

If all the points of such a set lie on one straight line, the set is called a linear set. In particular, if the straight line is the real axis, we have a set of real numbers. We presume that the reader is familiar, in general, with these as well as with plane point sets (*Elem.*, sec. III, ch. 6). He must also know the main features of the theory of infinite series, especially power series, and sequences of numbers (*Elem.*, sec. III, chs. 7 and 8). Many examples of these concepts are to be found in the chapters of the *Elem.* just mentioned. Every geometrical figure is a point set and every point set can be regarded as a geometrical figure.

The concepts of *greatest lower bound* and *least upper bound* in connection with sets of real numbers, and the theorem that every such set possesses a unique greatest lower bound as well as a unique least upper bound are particuarly important. Of course the theorem is valid in this generality only if the symbols $-\infty$ and $+\infty$ are also admitted as a greatest lower bound and a least upper bound, respectively. Otherwise it is only true if the set is "bounded on the left" or "bounded on the right." Equally important are the concepts of the *lower limit* and the *upper limit* ($\underline{\lim}$, $\overline{\lim}$, least and greatest limit point, respectively) of an infinite set of real numbers, and the theorem that these values are also uniquely determined by the set. Further details about sets of real numbers will not be discussed in this work.

Plane point sets may also be bounded or unbounded. A set $\mathfrak{M}$ is said to be *bounded* if all of its points can be enclosed in a figure of finite extent (e.g., in a circle). More precisely, the set is bounded if there exists a positive number K such that

$$|z| \leqq K$$

for all points z of the set. On the other hand, if there are points of $\mathfrak{M}$ outside of a circle of arbitrarily large radius with the center O, $\mathfrak{M}$ is said to be unbounded.

A point ζ of the plane is called a *limit point* of a set $\mathfrak{M}$ if an infinite number of points z of the set lie in every neighborhood of ζ (see §2, **h**)); in other words, if, for given (arbitrarily small) $\epsilon > 0$, there are always an infinite number of z for which

$$|z - \zeta| < \epsilon.$$

Numerous examples appear in *Elem.*, sec. III, ch. 6. The fundamental **Bolzano-Weierstrass theorem** (*Elem.*, §25) is concerned with such limit points:

Theorem 1. *Every bounded infinite (i.e., consisting of an infinite number of points) point set has at least one limit point.*

If the set is not bounded, this means, when referred to the sphere, that an infinite number of points of the set lie in every neighborhood (however small) of the north pole. In this case, we may call the point ∞ a limit point of the set. With this convention, the Bolzano-Weierstrass theorem holds for every infinite point set.

We recall, further, several simple concepts.

1. If $\mathfrak{M}$ is an arbitrary point set, then the points which do not belong to $\mathfrak{M}$ constitute the *complementary set* or *complement* of $\mathfrak{M}$. If all points of $\mathfrak{M}$ belong to another set $\mathfrak{N}$, then $\mathfrak{M}$ is called a *subset* of $\mathfrak{N}$.

2. If the defining property of a point set is such that no point having this property exists, the set is said to be "empty."

3. A point z_1 belonging to a set $\mathfrak{M}$ is called an "isolated point" of $\mathfrak{M}$ if there exists a neighborhood of z_1 containing no other point of the set.

4. A point z_1 belonging to a set $\mathfrak{M}$ is called an "interior point" of the set if there exists a neighborhood of z_1 belonging entirely to $\mathfrak{M}$.

5. A point ζ of the plane is called an "exterior point" with respect to a set $\mathfrak{M}$ if ζ itself and a neighborhood of it does not belong to $\mathfrak{M}$.

6. A point ζ of the plane is called a "boundary

point" of a set $\mathfrak{M}$ if there is at least one point which belongs to $\mathfrak{M}$ and at least one which does not belong to $\mathfrak{M}$ in every neighborhood of ζ. ζ itself may or may not belong to the set. According to this, an isolated point of a set $\mathfrak{M}$ or its complement is always a boundary point of $\mathfrak{M}$; it can never be an interior point.

7. A set is said to be "closed" if it contains all its limit points. The point ∞ is generally disregarded in this definition. Then it is more precise to say "closed in the plane"; otherwise, "closed on the sphere."

8. A set is said to be "open" if each of its points is an interior point of the set.

9. The least upper bound of the distances between two points of a set is called the "diameter" of the set. If the set is bounded and closed, then there are two points z_1, z_2 of the set such that its diameter is equal to $|z_2 - z_1|$; in short, the diameter is actually "assumed."

10. The greatest lower bound of the distances of a point ζ from the points of a set $\mathfrak{M}$ is called the "distance" of the point ζ from the set. If $\mathfrak{M}$ is closed, then there is a point z_0 in $\mathfrak{M}$ such that the distance of the point ζ from $\mathfrak{M}$ is equal to $|z_0 - \zeta|$; i.e., the distance is assumed.

11. The greatest lower bound of the distances $|z_1 - z_2|$ of a point z_1 of a set $\mathfrak{M}_1$ from a point z_2 of a set $\mathfrak{M}_2$ is called the "distance" between the two sets. If the sets are closed and if at least one of them is bounded, then the distance between them is assumed.

12. The "intersection" of two sets $\mathfrak{M}_1$ and $\mathfrak{M}_2$ is the set of all points which belong both to $\mathfrak{M}_1$ and $\mathfrak{M}_2$. Such an intersection may be empty (see 2). In that case $\mathfrak{M}_1$ and $\mathfrak{M}_2$ are called "disjunct" sets. A corresponding definition holds for any finite number or for an infinite number of sets.

13. The "logical sum" of two sets $\mathfrak{M}_1$ and $\mathfrak{M}_2$ is defined to be the set of all points which belong either to $\mathfrak{M}_1$ or to $\mathfrak{M}_2$. Again a corresponding definition holds for any finite number or for an infinite number of sets.

The *principle of nested intervals* (see *Elem.*, §27) now admits of a far-reaching generalization and leads to the so-called **theorem on nested sets**:

Theorem 2. *If $\mathfrak{M}_1$, $\mathfrak{M}_2, \ldots, \mathfrak{M}_n, \ldots$ is a sequence of entirely arbitrary closed point sets such that each is a subset of the preceding one, that at least one of the sets is bounded, and that their diameters tend to zero with increasing n, then there exists one and only one point ζ which belongs to all $\mathfrak{M}_n$.*

Proof: First it is clear that two distinct points ζ' and ζ'' cannot belong to all $\mathfrak{M}_n$; otherwise the diameters of all the sets would not be less than the fixed positive number $|\zeta'' - \zeta'|$, which is contrary to assumption. Then one notes that nearly all[1] sets are bounded; for nearly all the sets must have a finite diameter, and a set with a finite diameter is certainly bounded. Now, if a point is chosen from each set, say z_n from $\mathfrak{M}_n$, then the set of these z_n is bounded and therefore has a finite limit point ζ. This point belongs to all $\mathfrak{M}_n$; for if p is an arbitrary natural number, the sequence $z_p, z_{p+1}, \ldots$ is such that every element belongs to $\mathfrak{M}_p$. This sequence also has the limit point ζ. Since $\mathfrak{M}_p$ is closed, ζ also belongs to $\mathfrak{M}_p$ and hence to every one of the sets.

A theorem which is somewhat deeper and of great importance arises from the following circumstances. Every point z of a closed and bounded set $\mathfrak{M}$ is "covered" by a circle K_z; i.e., z lies in its interior. Consequently, a certain set (possibly infinite again) of circles exists such that every point of $\mathfrak{M}$ is covered by at least one of these circles.   (One and the same circle, however, may cover several points.) The **Heine-Borel theorem** then asserts the following:

Theorem 3. *If every point z of a closed and bounded set $\mathfrak{M}$ is covered by at least one circle K_z, then a finite number of these circles are sufficient to cover the set.*

Proof: We prove the theorem indirectly by showing

---

[1] That is, all except possibly a finite number (see *Elem.*, §26).

that the assumption that an infinite number of circles are necessary to cover $\mathfrak{M}$ contradicts the hypothesis that $\mathfrak{M}$ is closed. To this end, we first enclose the set $\mathfrak{M}$ in a square Q_1 and then divide Q_1 into four congruent subsquares. After annexing the sides, each of these four parts is closed. Then each of the four subsets of $\mathfrak{M}$ lying in one of the four subsquares is a closed and bounded set. Now assume that an infinite number of circles are necessary to cover the entire set; this must also be true for at least one of the four subsets. Call the first of the four squares[1] for which this is the case Q_2. From this one we obtain, in a similar manner, a square Q_3; and thus one finds a sequence of nested squares $Q_1, Q_2, \ldots, Q_n, \ldots$ (whose diameters decrease to zero) each of which contains a subset of $\mathfrak{M}$ requiring an infinite number of circles for its covering.

This cannot be the case, however, if $\mathfrak{M}$ is closed. For if the nest of squares shrinks to the point ζ, ζ is a limit point of $\mathfrak{M}$ and consequently belongs to $\mathfrak{M}$. Hence ζ is covered by one of the circles in question, say K_ζ. If p is chosen so large that the diagonal of Q_p is less than the distance of the point ζ from the circumference of the circle K_ζ, then all points of $\mathfrak{M}$ lying within Q_p are already covered by this one circle K_ζ; whereas an infinite number of circles were assumed to be necessary to cover these points. Since this is not the case, the theorem is true.

If a set is such that the numbers (points) which belong to it can be enumerated, i.e., designated in order as the first, second, $\ldots$, nth, $\ldots$ or as $z_1, z_2, \ldots, z_n, \ldots$, so that every element receives a definite number, then the set is called "enumerable." If this is not possible, the set is called "non-enumerable." (Cf. *Elem.*, ch. 7, where examples are also given.) If such an enumeration has been carried out, the set is said to be arranged in a *sequence of numbers (points)*. In

---

[1] We take the sides of all squares parallel to the coordinate axes and number the subsquares in the order in which the quadrants of the plane are usually numbered.

general, the same number is allowed to appear several times or even an infinite number of times in such a sequence. We then have the following general definition: If to every natural number $1, 2, \ldots, n, \ldots$ there corresponds, in an arbitrary manner, a single definite (complex) number $z_1, z_2, \ldots, z_n, \ldots$, respectively, then these numbers in the assigned order are said to form a *sequence of numbers*; and the points which correspond to them, a *sequence of points*. The sequence is designated briefly by $\{z_n\}$, and the single numbers z_n are called its "terms." Thus, we are simply concerned here with enumerable sets which have been enumerated (numbered throughout) in a certain definite manner, under the special agreement, however, that terms with different numbers need not necessarily be distinct. In the latter case, one and the same point is to be considered several or perhaps an infinite number of times as a point of the sequence: "it is counted several times or infinitely often." Hence, apart from this agreement, the effect of which is easily seen, the same considerations which have been carried through for arbitrary sets of numbers (points) hold for sequences of numbers (points). In particular, Theorems 1, 2, and 3 of this paragraph are valid; only it must be borne in mind that, on the basis of the agreement just made, a point ζ which appears infinitely often in a sequence of points is also a limit point of that sequence. ζ is said to be a limit point of the sequence $\{z_n\}$ if and only if, for a given (arbitrarily small) $\epsilon > 0$, an infinite number of z_n lie in the ϵ-neighborhood of ζ; i.e., if and only if

$$| z_n - \zeta | < \epsilon$$

for an infinite number of n. The case in which ζ is the only limit point of a sequence $\{z_n\}$ is of particular interest. The last relation then holds for all sufficiently large n, and consequently, for nearly all n (or all n "after a certain one," say for all $n > n_0 = n_0(\epsilon)$). ζ is called the "limit" of the sequence. We write

$$z_n \to \zeta \text{ for } n \to \infty \quad \text{or} \quad \lim_{n \to \infty} z_n = \zeta,$$

and the sequence of numbers $z_1, z_2, \ldots, z_n, \ldots$ is said to *converge to the limiting value* ζ.

Cauchy's general convergence principle furnishes a necessary and sufficient condition for this to occur (see *Elem.*, §26):

Theorem 4. *A necessary and sufficient condition for the sequence $z_1, z_2, \ldots, z_n, \ldots$ to have a limit is that for a given arbitrary $\epsilon > 0$ a number $n_0 = n_0(\epsilon)$ can be assigned such that*

$$| z_{n+p} - z_n | < \epsilon$$

for all $n > n_0(\epsilon)$ and all $p \geqq 0$.

If, from a given sequence of numbers $\{a_n\}$, a sequence of numbers $\{z_n\}$ is constructed by forming the sums

$$z_1 = a_1,\, z_2 = a_1 + a_2, \ldots,$$
$$z_n = (a_1 + a_2 + \ldots + a_n), \ldots$$

or the products

$$z_1 = a_1,\, z_2 = a_1 \cdot a_2, \ldots, z_n = (a_1 \cdot a_2 \ldots a_n), \ldots,$$

such a sequence is designated briefly by

$$\sum_{n=1}^{\infty} a_n,\ \prod_{n=1}^{\infty} a_n,$$

respectively. The first is called an "infinite series" with the terms a_n, the second, an "infinite product" with the factors a_n. The z_n are called the "partial sums" or the "partial products," in the respective cases. The reader is supposed to be familiar with the use of infinite series (see *Elem.*, chs. 7, 8).

Exercises. 1. Is the set defined by the relation

$$| z | + \Re(z) \leqq 1$$

bounded? Which part of the plane do the points of this set occupy?

2. Prove that every set consisting of isolated points only is enumerable.

3. Prove the assertions made in 10 and 11 that the distances mentioned there are "assumed."

4. Show that every limit point of a set $\mathfrak{M}$ which does not belong to that set is a boundary point of $\mathfrak{M}$, and every boundary point which does not belong to $\mathfrak{M}$ is a limit point of $\mathfrak{M}$.

5. Show that the totality of boundary points of a set is a closed set.

§4. Paths, Regions, Continua

In the following we frequently draw "paths" in the plane and consider "regions"; we must therefore give sharp definitions of these concepts.

1. If $x(t)$ and $y(t)$ are continuous (real) functions of t in the interval $\alpha \leqq t \leqq \beta$, then

$$x = x(t), \quad y = y(t)$$

is the parametric representation of a "continuous curve." If a continuous curve has no "multiple points," i.e., if two distinct points (x, y) correspond to two distinct values of t, it is called a *Jordan arc*. If one sets $x + iy = z$, so that $x(t) + iy(t) = z(t)$, then its representation can be written more briefly as

$$z = z(t), \quad \alpha \leqq t \leqq \beta.$$

$z(\alpha)$ is its initial point, $z(\beta)$ its terminal point. According to this, a Jordan arc is always "oriented"; i.e., it is always clear given two points on the arc, which precedes the other, and furthermore, which part of the arc is to be regarded as lying "between" them.

A *closed Jordan curve* is a continuous curve having $x(\alpha) = x(\beta)$, $y(\alpha) = y(\beta)$, but otherwise no multiple points.

A Jordan arc need not possess any assignable length. If it does have a definite length, the arc is said to be *rectifiable* and is then called a **"path segment."**

We cannot enter into a closer investigation of the concept of rectifiability here, but merely recall its definition. If the parameter interval $<\alpha, \ \beta>$ is divided in any manner into n parts, determined, say, by $\alpha = t_0 < t_1 < t_2 < \cdots < t_n = \beta$, and if the points

$z(t_\nu)$, $(\nu = 0, 1, 2, \ldots, n)$, are marked on the arc and joined in order by straight line segments, then an "inscribed segmental arc" is obtained. If the set of the lengths of all such inscribed segmental arcs is bounded, the arc in question is said to be *rectifiable*, and its length is defined as the least upper bound of that set. The Jordan arc given by the above parametric representation is rectifiable if and only if both functions $x(t)$ and $y(t)$ are of *bounded variation*. In particular, this is always the case if the derivatives $x'(t)$ and $y'(t)$ exist and are continuous in $<\alpha, \beta>$.

If a finite number of path segments are joined in order in such a manner that the initial point of each coincides with the terminal point of the preceding arc, a **"path"** is formed. A path, consequently, always possesses a definite length, is oriented, and admits of a representation of the form $z = z(t)$ such that as t runs over a certain (real) interval, the point z describes the entire path precisely once in a definite sense. The length of a path composed of several path segments is equal to the sum of the lengths of the single constituent segments, and correspondingly if a path is decomposed into several path segments by means of points of division. Unlike a path segment, a path may intersect itself in any manner. Because of the continuity of $x(t)$ and $y(t)$, the totality of points of a path is a *closed* point set.

If the initial and terminal points of a path coincide, it is called a **closed path.** It is oriented as before in the sense that $z(t)$ describes the entire closed path precisely once when t runs over its interval. If distinct points z always correspond in this manner to two distinct values of t, except the initial value and terminal value, the closed path is said to be *simple.* The following theorem concerns simple closed paths and, more generally, closed Jordan curves.

Jordan's Theorem. *A closed Jordan curve decomposes the plane into precisely two separated regions (see below), one lying inside and the other outside the curve.*

The proof of this important theorem, in spite of its apparent intuitive evidence, lies very deep and cannot be given here.[1] If the orientation of a simple closed path is such that the interior lies to the left, it is called positive orientation; otherwise, negative.[2] If nothing is said to the contrary, simple closed paths will be assumed to be oriented positively.

Every (oriented) straight line segment is, naturally, a path segment. If a finite number of straight line segments are joined in order in such a manner that the initial point of each coincides with the terminal point of the preceding segment, the resulting path is called a *segmental arc*. If its initial point and terminal point coincide, it is said to be closed or, more precisely, a *closed polygon*. If a closed polygon is simple, then, according to the last theorem, one can speak of its interior and its exterior.

We prove the following two lemmas for later application.

Lemma 1. *Every closed polygon p can be decomposed into a finite number of simple closed polygons and a finite number of segments described twice, once in each direction. Each of the former is described either entirely in the positive or entirely in the negative sense.*

Proof: Let us denote the sides

$$A_1A_2, \; A_2A_3, \; \ldots, \; A_{n-1}A_n, \; A_nA_{n+1}$$

of p by

$$s_1, \; s_2, \; \ldots, \; s_{n-1}, \; s_n,$$

respectively; here $n \geqq 2$, A_1 is the initial point and A_{n+1} is the terminal point of p, and $A_1 = A_{n+1}$. We may suppose, without loss of generality, that no two successive sides are such that they have only one point in common and lie on the same straight line. In the following discussion s_n is assumed to be a segment which is open at A_{n+1}.

---

[1] See G. N. Watson. *Complex Integration and Cauchy's Theorem,* Cambridge Tracts No. 15, 1914. ch I for a proof

[2] For a more precise definition of positive orientation see *op. cit.,* pp. 15, 16.

One and only one of the following is true.

1) *Every*

$$s_\nu \qquad (\nu = 2, 3, \ldots, n)$$

has only one point in common with $s_{\nu-1}$ and no point in common with

$$s_\mu \qquad (\mu = 1, 2, \ldots, \nu - 2).$$

In this case p is simple and there is nothing further to prove.

2) *There exists an s_ν, with $\nu \geqq 2$, such that*

    a) *s_ν has more than one point in common with $s_{\nu-1}$,* or

    b) *s_ν has at least one point in common with one or more of the segments*

$$s_\mu \qquad (\mu = 1, 2, \ldots, \nu - 2).$$

In this case let s_k be the first such s_ν (i.e., the s_ν with smallest subscript).

If a) holds for s_k, either there is a point B_{k-1} of s_{k-1} such that

$$B_{k-1} = A_{k+1},$$

and then $B_{k-1}A_kA_{k+1}$, described in that order, is a straight line segment q' described twice, once in each direction, and $\overline{A_1A_2}\overline{A_2A_3} \cdots \overline{A_{k-1}B_{k-1}}\overline{A_{k+1}A_{k+2}} \cdots \overline{A_nA_{n+1}}$ is a closed polygon p'; or there is a point B_k of s_k such that

$$A_{k-1} = B_k,$$

and then $A_{k-1}A_kB_k$, described in that order, is a straight line segment q' described twice, once in each direction, and $\overline{A_1A_2}\overline{A_2A_3} \cdots \overline{A_{k-2}A_{k-1}}\overline{B_kA_{k+1}} \cdots \overline{A_nA_{n+1}}$ is a closed polygon p'. ($B_kA_kA_{k+1}$ and $A_{k-1}A_kB_k$ are considered degenerate forms of a closed polygon.)

If a) does not hold for s_k but b) does, let B_k on A_kA_{k+1} be the nearest point to A_k that s_k has in common with any of the segments

$$s_\mu \qquad (\mu = 1, 2, \ldots, k - 2),$$

and let

$$B_r = B_k,$$

where B_r is on s_r for some $r \leqq k - 2$. There can be only one such B_r because of the way in which s_k was chosen. Then

$$\overline{B_r A_{r+1}} \overline{A_{r+1} A_{r+2}} \cdots \overline{A_k B_k}$$

is a simple closed polygon q'. For, due to the manner in which s_k was selected, if q' were not simple, $A_k B_k$ would have to have a point distinct from B_k in common with some preceding segment; but this is impossible by the definition of B_k. q' is described in the sense of the orientation of p and hence, since q' is simple, either entirely in the positive or entirely in the negative sense. $\overline{A_1 A_2} \overline{A_2 A_3} \cdots \overline{A_r B_r} \overline{B_k A_{k+1}} \cdots \overline{A_n A_{n+1}}$ is a closed polygon p'.

In either case, 1) or 2), p is thus decomposed into a simple closed polygon q' (or a segment described twice, once in each direction) and a closed polygon p'. If p' is simple, our proof is complete; if not, then the above argument applied to p' will lead to a decomposition of the latter into a simple closed polygon q'' (or segment described twice) and a closed polygon p''. It is clear that by continuing in this manner we obtain after a finite number of steps the decomposition stated in the lemma, because every side of p can have only a finite number of maximal subsegments in common with the other sides, and only a finite number of points not belonging to such subsegments in common with the other sides.

Lemma 2. *Every simple closed polygon can be decomposed into triangles by means of diagonals lying in the interior of the polygon.*

We prove this by induction on the number of vertices of the polygon. The lemma is obviously true for quadrilaterals (with or without re-entering angles; see Fig. 2, p. 52). Let p be a polygon with n (>4) vertices, and assume that the lemma has been proved for polygons with fewer than n vertices. Then it suffices to show the existence of an interior diagonal which decomposes p

into two subpolygons; for, each of the latter then has fewer than n vertices. This can be done as follows. Let a straight line which does not intersect p be translated parallel to itself toward the polygon until they meet. Then the line necessarily contains a vertex A of p, and the interior angle of the polygon at A is less than two right angles. Let B and C denote the vertices adjacent to A. Then precisely one of the following is true:

1) BC is a diagonal lying in the interior of p;

2) there is at least one vertex of p on the (open) segment BC (let one of these vertices be denoted by V), but no vertex in the interior of triangle ABC;

3) there is at least one vertex of p in the interior of triangle ABC. If 1) is true, then there is nothing further to show. If 2) holds, then AV is an interior diagonal of p. If 3) is true, let a point X move from B to C along BC until AX encounters a vertex or vertices of p in the interior of the triangle ABC. If V denotes that one of these vertices which is nearest to A, then AV is a diagonal in the interior of p.[1]

2. Every point set which

a) contains only *interior* points, and is therefore *open* (see §3, 8), and which is

b) *connected*

is called a **region.**

An *open* point set is said to be *connected* if any two of its points can be joined by a segmental arc belonging entirely to this point set.

According to this definition, in speaking of a *region* we do **not** include its boundary points. A *region together with its boundary points* will always be referred to as a *closed region.*

Regions can assume very many different forms. For example, besides such simple regions as the circle, polygon, half-plane, the point set consisting of the upper half-plane, $\Im(z) > 0$, with the omission of all points

---

[1] For a more vigorous treatment of this lemma, see N. J. Lennes, Amer. J. Math., 33 (1911), pp. 45–47.

lying on the perpendiculars of unit length erected upon the real axis at the points 0 and $\pm \dfrac{1}{n}$, $(n = 1, 2, \ldots)$, is a region. Observe that the boundary point 0 cannot be reached along any path lying wholly within this region.

Special attention is called to those regions which are *simply connected*. A region is said to be *simply connected* if every simple closed path lying entirely within the region encompasses only points of the region itself (and consequently, no boundary points).

The circle, triangle, interior of a closed Jordan curve are simply connected. On the other hand, the region between two concentric circles is not simply connected, nor is the region $|z| > 0$.

For later use we need also the following:

Lemma 3. *If a path k (or more generally, a closed point set) lies within a region $\mathfrak{G}$, then there is a positive number ρ such that the distance of every point of the path from the boundary of the region is greater than ρ; i.e., the path k does not come arbitrarily close to the boundary.*

Proof: Since every point z of k lies in $\mathfrak{G}$, a circular neighborhood about z as center with radius ρ_z, say, also belongs entirely to $\mathfrak{G}$. Now, as in the Heine-Borel theorem, let there correspond to each of these points z the circle with center z and radius $\frac{1}{2}\rho_z$. Then according to this theorem, a finite number of these circles are sufficient to cover k. Let ρ be the radius of the smallest of these. Then ρ satisfies the conditions of the lemma, since a circle of radius ρ certainly lies entirely within $\mathfrak{G}$, even if its center lies on the circumference of one of that finite number of covering circles.

3. Every *bounded* point set which is
a) *closed* and
b) *connected*
is called a **continuum**.
A *closed* and bounded point set is said to be *connected*

if any two of its points A and B can be joined by means of an "ϵ-chain," that is to say, if for given $\epsilon > 0$ a finite number of points of the set, say $A_0 = A$, A_1, A_2, ..., $A_n = B$, can be assigned so that the distance between any two consecutive points is less than ϵ.

Since continua can have the most varied forms, it is often useful to be able to replace them by simpler configurations. In this connection we state the following lemma, whose proof is omitted because (like the proof of Jordan's theorem) it raises difficulties in its complete generality. On the other hand, it is almost self-evident for simple sets.

Lemma 4. *If K is a continuum, then the complement of K is composed of one or more regions. Precisely one of them, call it $\mathfrak{G}$, contains arbitrarily distant points of the plane. $\mathfrak{G}$ is called the exterior region determined by K. If $\epsilon > 0$ is chosen arbitrarily, there always exists a simple closed polygon P belonging entirely to $\mathfrak{G}$ (so that K therefore lies in the interior of P) such that the distance of every point of P from K is less than ϵ.*

CHAPTER 2

FUNCTIONS OF A COMPLEX VARIABLE

§5. The Concept of a Most General (Single-valued) Function of a Complex Variable

If $\mathfrak{M}$ is an arbitrary point set and if z is allowed to denote any point of $\mathfrak{M}$, z is called a (complex) *variable* and $\mathfrak{M}$ is called the *domain of variation* of z.

If there is a rule by means of which a definite new number w is made to correspond to every point z of $\mathfrak{M}$, w is called a *(single-valued) function* of the (complex) variable z; in symbols

$$w = f(z),$$

where "f" stands for the prescribed rule. $\mathfrak{M}$ is called the "domain of definition" and z the "argument" of the function. The totality of values w which correspond to the points z of $\mathfrak{M}$ is called the "domain of values" of the function (over $\mathfrak{M}$). Any other symbols may be employed instead of f; F, g, h, φ, etc. will often be used.

If z and w are separated into their real and imaginary parts, $z = x + iy$, $w = u + iv$, then the relation

$$w = f(z)$$

can also be interpreted to mean that to the pair of real numbers x and y there correspond, by means of certain rules, two new real numbers u and v. Thus, u and v appear as a pair of real functions of two real variables, x and y. We set

$$u = u(x, y), v = v(x, y),$$

and consequently

$$f(z) = u(x, y) + iv(x, y).$$

21

u is called the *real part*, and v, the *imaginary part* of the function $f(z)$. According to this, it is evident that $f(z)$ is merely a combination of a pair of real functions of two real variables. It is sometimes useful to place this interpretation in the foreground; this will be done, e.g., in §§7 and 10. In general, however, the real core of the matter can be perceived only if this separation does not take place and $f(z)$ is considered as a function of the *single* complex variable z.

We presume that the reader is already familiar, to some extent, with the so-called *elementary functions*, including the rational functions (particularly the linear functions), the exponential function e^z, the trigonometric functions $\sin z$, $\cos z$, $\tan z$, and their inverses (see *Elem.*, secs. II and V). For these functions, $\mathfrak{M}$ is either the entire plane, as for e^z, $\sin z$, $\cos z$, or the plane with the exception of certain points; e.g., for the rational functions, the zeros of the denominator are excluded; for $\cot z$, all real points of the form $k\pi$, $k = 0$, $\pm 1, \ldots$ are excluded. Here the rule for defining the function consists in an explicit expression; i.e., the value w of the function corresponding to a z of $\mathfrak{M}$ can be calculated by means of a finite or an infinite[1] number of applications of the four fundamental operations of arithmetic.

The prescribed rule, however, can be given in an entirely different manner. Only to mention an extreme example, let $\mathfrak{M}$ be the set of all numbers $z = x + iy$ for which x and y are rational numbers, and stipulate that $f(z)$ is equal to $1, 2, \ldots$, or n according as the periodic decimal expansion of y has a period of $1, 2, \ldots$, or n digits, respectively.

It should be emphasized immediately that it is by no means necessary for a function to be given by an explicit expression. It can be given in very many other ways; all that is required is that the value w of

---

[1] In this case, the limit process in question, usually infinite power series, must, naturally, converge.

the function be made to correspond, on the basis of the definition, to each z of $\mathfrak{M}$ in a completely unambiguous manner.   It is evident that the concept of function thus formulated is exceedingly broad, so broad that it can hardly be governed by general theorems and rules. It will be our task to restrict the hypotheses in a suitable manner in order to select from the totality of all functions a more special class of functions which are valuable with regard to their applicability in mathematics and the physical sciences.

It is surprising that this objective is attained with the single and quite natural requirement that our functions be *differentiable*.   It is also surprising that the property of being differentiable has unexpected, far-reaching consequences for the nature of the function.

Differentiability, which is defined formally the same as in the real domain, likewise presupposes *continuity*. We also regard these two concepts and their simplest properties as being familiar to the reader (see *Elem.*, sec. IV).   The most important facts concerning them appear in the following paragraph.

§6. Continuity and Differentiability

I. We first require that the domain of variation $\mathfrak{R}$ be a region $\mathfrak{G}$ in the sense of §4, 2.[1] z, then, is said to be a continuous variable; for if ζ is any point of $\mathfrak{G}$, z may represent every point of a neighborhood of ζ, and hence, every point sufficiently close to ζ.   A function $w = f(z)$ defined in $\mathfrak{G}$ is said to be continuous at a point ζ of $\mathfrak{G}$ if it satisfies one of the following fully equivalent conditions, (formally the same as in the real domain).

FIRST FORM. $\lim\limits_{z \to \zeta} f(z)$ exists and is equal to $f(\zeta)$; that is, having chosen $\epsilon > 0$, it is always possible to

---

[1] In what follows, it usually suffices to think of $\mathfrak{G}$ as representing the interior of a circle.

assign a number $\delta = \delta(\epsilon) > 0$ such that, with $\omega = f(\zeta)$,

$$| w - \omega | = | f(z) - f(\zeta) | < \epsilon$$

for all z for which

$$| z - \zeta | < \delta.$$

This can also be said in the following, less precise manner.

SECOND FORM. The values $f(z)$ of the function differ from $f(\zeta)$ by arbitrarily small amounts when z lies sufficiently close to ζ.

THIRD FORM. If an entirely arbitrary sequence of numbers, $z_1, z_2, \ldots, z_n, \ldots$, of $\mathfrak{G}$ is chosen such that $z_n \to \zeta$, then for the corresponding values $w_n = f(z_n)$ of the function

$$w_n \to \omega = f(\zeta).$$

If a function $f(z)$ is continuous at every point of a region, then it is said to be *continuous in the region*.

Occasionally the functions which occur are also defined for some boundary point of $\mathfrak{G}$. Then the continuity of the function $f(z)$ at a boundary point ζ of $\mathfrak{G}$ is understood to mean that the conditions for continuity are fulfilled at least if the z which appear in them lie within $\mathfrak{G}$. In this sense one speaks of "continuity from the interior." Similarly, one also speaks of "continuity along a path," which means that the conditions for continuity are fulfilled for all points lying on the path in question, irrespective of the values of the function for other points.

If it is possible to make a value $\omega = f(\zeta)$ correspond to a boundary point ζ of the region of definition $\mathfrak{G}$ in such a way that $f(z)$ is now continuous at ζ from the interior, even if it is necessary to alter an already defined value of the function for ζ, the function $f(z)$ is said to *assume the boundary value* ω at ζ. This is obviously the case if and only if lim $f(z)$ exists for z approaching ζ from the interior of $\mathfrak{G}$.

The continuity of $f(z)$ evidently implies that the functions $u(x, y)$ and $v(x, y)$ introduced in the preceding paragraph must, for their part, be continuous, real functions of the pair of variables (x, y).

As for these functions, the following **theorem on uniform continuity** also holds for our continuous functions of a complex variable.

Theorem. *If $f(z)$ is continuous in a closed and bounded region $\overline{\mathfrak{G}}$, then, having chosen $\epsilon > 0$, it is always possible to assign a number $\delta = \delta(\epsilon)$ in such a manner, that for any two points z' and z'' of $\overline{\mathfrak{G}}$ for which $| z'' - z' | < \delta$, the modulus of the difference of the corresponding values of the function*

$$| w'' - w' | = | f(z'') - f(z') | < \epsilon.$$

Proof: A circle, whose radius we denote by ρ_z, can be drawn about every point z of $\mathfrak{G}$ as center such that the oscillation[1] of the function in that circle is less than $\frac{1}{2}\epsilon$, because of the continuity of $f(z)$ at z. Now, to every z of $\overline{\mathfrak{G}}$ we let correspond, as in the proof of Lemma 3, §4, the circle about z as center with radius $\frac{1}{2}\rho_z$. By the Heine-Borel theorem, a finite number of these circles are sufficient to cover $\overline{\mathfrak{G}}$. If the radius of the smallest of these circles is δ, this number satisfies the conditions of the theorem. For, if $| z'' - z' | < \delta$ and if z' is covered, say, by the circle about ζ as center with radius $\frac{1}{2}\rho_\zeta$, then $\delta \leq \frac{1}{2}\rho_\zeta$; and consequently z' and z'' lie within the circle about ζ as center with radius ρ_ζ. Hence $| f(z'') - f(z') | < \epsilon$.

II. The definition of *differentiability*, which is also formally the same as in the real domain, will likewise be stated in three different forms. A function $w = f(z)$ defined in $\mathfrak{G}$ is said to be differentiable at a point ζ of $\mathfrak{G}$

---

[1] That is, the least upper bound of the values $|f(z'') - f(z')|$ for any two points z' and z'' of the circle in question which also lie in $\overline{\mathfrak{G}}$.

if one of the following three equivalent conditions is satisfied.

FIRST FORM.

$$\lim_{z \to \zeta} \frac{f(z) - f(\zeta)}{z - \zeta}$$

exists. This limit is denoted by $f'(\zeta)$ or $(dw/dz)_{z=\zeta}$ and is called the *derivative* or *differential quotient* of $f(z)$ at the point ζ. In other words, it must be possible to associate a new number $f'(\zeta)$ with the point ζ in such a way that having chosen $\epsilon > 0$ arbitrarily, a $\delta = \delta(\epsilon)$ can always be found such that

$$\left| \frac{f(z) - f(\zeta)}{z - \zeta} - f'(\zeta) \right| < \epsilon$$

for all z of $\mathfrak{G}$ with $|z - \zeta| < \epsilon$. This can be said (somewhat less precisely) as follows:

SECOND FORM. For all z of $\mathfrak{G}$ lying sufficiently close to ζ, the *difference quotient*

$$\frac{f(z) - f(\zeta)}{z - \zeta} = \frac{w - \omega}{z - \zeta} = \left(\frac{\Delta w}{\Delta z} \right)_{z=\zeta}$$

lies arbitrarily close to a definite number, which number is then denoted by $f'(\zeta)$.

THIRD FORM. If an entirely arbitrary sequence of numbers, $z_1, z_2, \ldots, z_n, \ldots$, of $\mathfrak{G}$ is chosen, whose terms all differ from ζ but approach ζ as a limit, then the sequence of numbers

$$\Delta_n = \frac{f(z_n) - f(\zeta)}{z_n - \zeta}$$

always tends to a limit. The latter is independent of the choice of the sequence $\{z_n\}$ and is denoted by $f'(\zeta)$.

We assume that the rules of differentiation, formally the same as those in the real domain, and, in particular, the so-called "chain rule" are familiar to the reader (see *Elem.*, sec. IV, ch. 9). Likewise, the meaning of continuity and differentiability in connection with the interpretation of a function $w = f(z)$ as a mapping of the region of definition in the z-plane onto a region in the w-plane is assumed to be known. In a few words, continuity means that neighboring points in the z-plane correspond to neighboring points in the w-plane, and differentiability means that the mapping is *conformal*[1] (see *Elem.*, sec. IV, ch. 10).

A function which is differentiable at every point of a region is said to be *differentiable in the region*. The derivative then is also a function defined in this region. Those functions which are differentiable in regions are the ones which were alluded to in the preceding paragraph and which will prove to be very important. They are therefore given a special name.

Definition. *A function which is defined and differentiable throughout a region $\mathfrak{G}$ is called a (single-valued)* **regular analytic** *function in $\mathfrak{G}$, or briefly, an analytic or a regular function. The region $\mathfrak{G}$ is called a region of regularity of the function.*

According to this, the property of being regular belongs to a function only in *regions;* however, the function is also said to be regular at every single point of such a region. Note then that regularity at a point always automatically includes regularity in a certain neighborhood thereof, since this point *eo ipso* must be an **interior** point of a region of regularity. All the elementary functions mentioned above are regular in their regions of definition. The function $f(z) = \Re(z)$ is easily seen to be a function which is continuous in the entire plane but not a regular analytic function in any region.

The succeeding sections will bear out the fact that every member of the class of functions thus selected possesses a surprisingly strong inner structure. These

---

[1] Angles are preserved in magnitude and sense, and magnification at a point is independent of direction.

functions, therefore, are especially important for all applications in the mathematical sciences.

Exercises. 1. Investigate the continuity of the following two functions:

α) $f(z) = 0$ for $z = 0$ and for all z whose absolute value $|z|$ is an irrational number;

$$f(z) = \frac{1}{q} \qquad \text{if} \quad |z| = \frac{p}{q},$$

where p and q are positive and relatively prime integers.

β) $f(z) = 0$ for $z = 0$, $f(z) = \sin \theta$ for $z = r(\cos \theta + i \sin \theta)$ with $r > 0$.
For both functions, determine the points at which they are continuous and at which they are discontinuous.

2. Are the functions defined in the previous exercise differentiable at certain points? Are the functions $f(z) = |z|$, $f(z) = \Re(z)$, $f(z) = $ am z differentiable at certain points?

3. Let the function $f(z)$ be continuous in a circle K (more generally, in the interior of a simple closed path C) and assume a boundary value $f(\zeta)$ for every boundary point ζ. Show that these boundary values $f(\zeta)$ form a continuous function along K (or C).

§7. The Cauchy-Riemann Differential Equations

The significance, as far as the functions $u(x, y)$ and $v(x, y)$ are concerned, of the requirement that $f(z) = u + iv$ be differentiable at the point $\zeta = \xi + i\eta$ can be realized as follows. The difference quotient

$$\left(\frac{\Delta w}{\Delta z}\right)_{z=\zeta} = \frac{[u(x, y) + iv(x,y)] - [u(\xi, \eta) + iv(\xi, \eta)]}{(x + iy) - (\xi + i\eta)}$$

must always tend to a single definite number as a limit, howsoever $z \rightarrow \zeta$. In particular, the limit must exist if z is allowed to approach ζ once along a line parallel to the x-axis and another time along a line parallel to the y-axis; that is, if for fixed $y = \eta$, x is made to approach ξ, and if for fixed $x = \xi$, y is made to approach η. Thus, we have the following result.

Theorem 1. *If the function $f(z) = u(x, y) + iv(x, y)$ is differentiable at the point $\zeta = \xi + i\eta$, then the four*

*partial derivatives of u and v with respect to ξ and η
exist there:*

$$\frac{\partial u}{\partial x} = u_x(\xi, \eta), \frac{\partial u}{\partial y} = u_y(\xi, \eta), \frac{\partial v}{\partial x} = v_x(\xi, \eta), \frac{\partial v}{\partial y} = v_y(\xi, \eta).$$

Then for the two methods in which $z \to \zeta$,

(1) $f'(\zeta) = u_x + iv_x,\ f'(\zeta) = \dfrac{1}{i}(u_y + iv_y),$

respectively. From this we obtain the following
theorem which, as in the real domain, is of fundamental
importance for the integral calculus.

Theorem 2. *If a function $f(z)$ is differentiable in a
region $\mathfrak{G}$, and if its derivative is zero everywhere in $\mathfrak{G}$,
then $f(z) \equiv c$ in $\mathfrak{G}$; or, two functions which are regular
in the same region $\mathfrak{G}$ and whose derivatives coincide there
differ in $\mathfrak{G}$ only by an additive constant.*

For, both partial derivatives of u and likewise those
of v are zero everywhere in $\mathfrak{G}$. Hence, u, v, and consequently also $f(z)$ are identically constant in $\mathfrak{G}$.

Since the two values in (1) must be equal, we also
obtain the following theorem.

Theorem 3. *If the function $f(z) = u + iv$ is differentiable at the point $\zeta = \xi + i\eta$, then the relations*

$$\frac{\partial u}{\partial x} = \frac{\partial v}{\partial y}, \qquad \frac{\partial u}{\partial y} = -\frac{\partial v}{\partial x},$$

*involving the four partial derivatives of u and v, hold at
the point (ξ, η). They hold then, in particular, at every
point of a region of regularity of $f(z)$.*

These important equations, which must be satisfied
by the real part and the imaginary part of $f(z)$, are
called the **Cauchy-Riemann** (*partial*) **differential equations.** The importance of these differential equations
depends on the fact that they are characteristic for
regular functions; for, the converse of Theorem 3 is
also true.

Theorem 4. *If the four partial derivatives of u and v with respect to x and y exist in a region of the z- or xy-plane, and if they are continuous and satisfy the Cauchy-Riemann differential equations, then*

$$f(z) = u(x, y) + iv(x, y)$$

is a regular function of z in $\mathfrak{G}$.

Proof: We have

$$f(z) - f(\zeta) = [u(x,y) + iv(x,y)] - [u(\xi, \eta) + iv(\xi, \eta)].$$

By the theorem on the total differential for real functions of two real variables we may write

$$u(x, y) - u(\xi, \eta) =$$
$$[u_x(\xi, \eta) + \alpha(x, y)](x - \xi) + [u_y(\xi, \eta) + \beta(x, y)](y - \eta)$$

and

$$v(x, y) - v(\xi, \eta) =$$
$$[v_x(\xi, \eta) + \gamma(x, y)](x - \xi) + [v_y(\xi, \eta) + \delta(x, y)](y - \eta),$$

where α, β, γ, δ denote functions of x and y which tend to zero as $(x, y) \to (\xi, \eta)$.

Since obviously $\left| \dfrac{x - \xi}{z - \zeta} \right| \leqq 1$ and $\left| \dfrac{y - \eta}{z - \zeta} \right| \leqq 1$, we immediately infer from the last two equations, bearing in mind the Cauchy-Riemann differential equations, that

$$\frac{f(z) - f(\zeta)}{z - \zeta} \to u_x(\xi, \eta) + iv_x(\xi, \eta)$$

as $z \to \zeta$. Therefore $f(z)$ is differentiable at the point ζ, and hence, everywhere in $\mathfrak{G}$.

Thus, the Cauchy-Riemann equations characterize in a unique manner those functions of the form $u(x, y)$ and $v(x, y)$ which can be the components of an analytic function.

If we assume further the existence and continuity in $\mathfrak{G}$ of the second-order partial derivatives (it will be proved in §16 that this is always automatically the case), then it follows from the Cauchy-Riemann equations that

$$\frac{\partial^2 u}{\partial x^2} = \frac{\partial^2 v}{\partial y \partial x} \quad \text{and} \quad \frac{\partial^2 u}{\partial y^2} = -\frac{\partial^2 v}{\partial x \partial y} = -\frac{\partial^2 v}{\partial y \partial x}.$$

Hence

$$\frac{\partial^2 u}{\partial x^2} + \frac{\partial^2 u}{\partial y^2} = 0,$$

and likewise

$$\frac{\partial^2 v}{\partial x^2} + \frac{\partial^2 v}{\partial y^2} = 0.$$

Both functions u and v satisfy one and the same differential equation, **Laplace's differential equation,** as it is called, of the form

$$\frac{\partial^2 \varphi}{\partial x^2} + \frac{\partial^2 \varphi}{\partial y^2} = 0.$$

It follows that neither the real nor the imaginary part of $f(z)$ can be chosen arbitrarily; on the contrary, each alone must satisfy Laplace's equation, and both together must satisfy the Cauchy-Riemann equations.

Exercises. 1. Show that the Cauchy-Riemann equations and Laplace's equation are satisfied by the elementary functions, e.g., by

$$f(z) = z, \ z^2, \ z^n, \ e^z, \ \sin z, \ \cos z, \ \tan z, \text{ etc.}$$

2. Prove Theorem 2 of this paragraph without resolving $f(z)$ into its real and imaginary parts.

INTEGRAL THEOREMS

CHAPTER 3

THE INTEGRAL OF A CONTINUOUS FUNCTION

§8. Definition of the Definite Integral

In the integral calculus, the definite integral of a real continuous function $y = F(x)$ of the real variable x, taken between the limits x_0 and X, is defined as follows:

Divide the interval $< x_0, X >$ (take $x_0 < X$) in any manner into n parts. Let the points of division be

$$x_0 < x_1 < x_2 < \cdots < x_{n-1} < x_n = X.$$

In each interval $< x_{\nu-1}, x_\nu >$ choose an arbitrary point ξ_ν and form the sum

$$J_n = \sum_{\nu=1}^{n} (x_\nu - x_{\nu-1})F(\xi_\nu).$$

Let this be carried out for $n = 1, 2, 3, \ldots$, each time in an entirely arbitrary manner, but so that the lengths of *all* intervals $< x_{\nu-1}, x_\nu >$ decrease to zero with increasing n. Then

$$\lim_{n \to \infty} J_n = J$$

always exists and is completely independent of the choice of the points of division and the intermediate points. In other words, a number J exists such that for given $\epsilon > 0$ there exists a $\delta = \delta(\epsilon) > 0$ such that

$$| J_n - J | < \epsilon,$$

32

provided all intervals

$$| x_\nu - x_{\nu-1} | < \delta.$$

This number J is called the definite integral and is denoted by

$$J = \int_{x_0}^{X} F(x)dx.$$

We presume that the reader is familiar with this definition of the real definite integral and its geometrical interpretation as the approximation of a plane area by means of a sum of rectangles.

Now let $w = f(z)$ be a continuous function of z in a region $\mathfrak{G}$ (differentiability is not necessary for the present). Let z_0 and Z be two arbitrary points of $\mathfrak{G}$. The following definition of the definite integral of a function of a complex variable is formally analogous to the one given above. Connect z_0 and Z by means of a path k lying entirely within $\mathfrak{G}$. Divide k into n parts in any manner. Call the points of division, in order, $z_0, z_1, z_2, \ldots, z_{n-1}, z_n = Z$. On each of the paths $z_{\nu-1} \ldots z_\nu$ choose an arbitrary point ζ_ν and form the sum

$$J_n = \sum_{\nu=1}^{n} (z_\nu - z_{\nu-1})f(\zeta_\nu).$$

We shall show that in this case too

$$\lim_{n \to \infty} J_n = J$$

always exists and is independent of the choice of the points of division and of the intermediate points, provided the lengths of all paths $z_{\nu-1} \ldots z_\nu$ decrease to zero with increasing n. This limit is not independent of the connecting path k, however. Thus, we shall prove the existence of a number J with the following property. For given $\epsilon > 0$, a $\delta = \delta(\epsilon) > 0$ can be determined such that

$$| J_n - J | < \epsilon,$$

provided the lengths of all paths $z_{\nu-1} \ldots z_\nu$ are less than δ.

The limiting value

$$J = \lim_{n \to \infty} \left\{ \sum_{\nu=1}^{n} (z_\nu - z_{\nu-1}) \, f(\varsigma_\nu) \right\},$$

understood in this sense, is called the definite integral of $f(z)$ taken along k and is denoted by

$$\int_{z_0}^{Z}{}_{\!\!k} f(z)dz \quad \text{or, briefly, by} \quad \int_k f(z)dz.$$

A simple geometrical interpretation as in the case of real integrals is impossible.

§9. Existence Theorem for the Definite Integral

For brevity we shall call the sums in question Σ-sums (of n parts); and when we speak of a segment (a, b) of the path, we shall always mention first that point which precedes on the oriented path. With these conventions we have

Lemma 1. *Let (a, b) be a segment k', of length l', of the path k. Let the oscillation of the function $f(z)$ on k'[1] be less than σ. Then two Σ-sums which are formed for this segment for $n = 1$ and $n = p$ $(\geqq 1)$ differ by an amount less than $l'\sigma$.*

Proof: Let $s = (b - a)f(\alpha_0)$ and $s' = (a_1 - a)f(\alpha_1) + (a_2 - a_1)f(\alpha_2) + \cdots + (b - a_{p-1})f(\alpha_p)$ be the two Σ-sums. Here we have denoted the points of division

---

[1] That is, the least upper bound of the values $|f(z') - f(z'')|$ for any two points z' and z'' of the segment k'.

by $a_1, a_2, \ldots, a_{p-1}$ and the intermediate points by α_0, $\alpha_1, \alpha_2, \ldots, \alpha_p$, respectively. By hypothesis,

$$| f(\alpha_\nu) - f(\alpha_0) | < \sigma, \quad \nu = 1, 2, \ldots, p.$$

Since s can be written in the form

$$s = (a_1 - a)f(\alpha_0) + (a_2 - a_1)f(\alpha_0) + \cdots + (b - a_{p-1})f(\alpha_0),$$

we have

$$| s' - s | < \sigma(| a_1 - a | + | a_2 - a_1 | + \cdots + | b - a_{p-1} |) \leqq l'\sigma,$$

because the length of an inscribed segmental are (cf. §4, 1) cannot be greater than the length of the path itself.

Lemma 2. *Let S be a fixed Σ-sum, of n parts say, for the path k, and let the oscillations of $f(z)$ on the n segments of the path all be less than σ_0. Let S' be a new Σ-sum, derived from S by adding new points of division to the old ones (briefly, by further subdivision). Then, if l denotes the length of the path k, we have*

$$| S - S' | < l\sigma_0,$$

no matter how the intermediate points defining S' are chosen.

Proof: Lemma 1 holds for each of the n segments of the path, so that we have

$$| S - S' | < l_1\sigma_0 + l_2\sigma_0 + \cdots + l_n\sigma_0 = l\sigma_0,$$

if $l_1, l_2, \ldots, l_n$ denote the lengths of the n segments of the path.

Lemma 3. *Given $\epsilon > 0$, there exists a $\delta = \delta(\epsilon) > 0$ such that if S_1 and S_2 are any two Σ-sums defined by means of decompositions of the path into segments of lengths less than δ, then*

$$\left| S_1 - S_2 \right| < \frac{\epsilon}{2}.$$

Proof: Choose δ so that $|f(z'') - f(z')| < \dfrac{\epsilon}{4l}$ for any two points z' and z'' of the path for which $|z'' - z'| < \delta$. This is possible by virtue of the theorem on uniform continuity. If S_1 and S_2 are any two Σ-sums for whose decompositions all segments of the path have lengths less than δ, form a third (finer) decomposition by taking as points of division those of the first two decompositions. The third is evidently derived from them by means of *further subdivision.* Hence, if S_3 is an arbitrary Σ-sum belonging to the third decomposition, we have by Lemma 2

$$|S_1 - S_3| < l\cdot\frac{\epsilon}{4l} = \frac{\epsilon}{4},$$

and likewise

$$|S_2 - S_3| < \tfrac{1}{4}\epsilon.$$

Therefore,

$$|S_1 - S_2| = |(S_1 - S_3) - (S_2 - S_3)|$$
$$\leqq |S_1 - S_3| + |S_2 - S_3| < \tfrac{1}{2}\epsilon,$$

Q. E. D.

Lemma 4. *Let a Σ-sum be formed for $n = 1, 2, \ldots$. If the lengths of all the path segments of the respective decomposition decrease to zero with increasing n[1], then*

$$\lim_{n\to\infty} S_n$$

exists.

Proof: Given $\epsilon > 0$, determine δ according to Lemma 3. Then take n_0 so large, that the lengths of all segments of all the S_n with $n \geqq n_0$ are less than δ. Lemma 3 is applicable to all these S_n; i.e.,

$$|S_{n+p} - S_n| < \tfrac{1}{2}\epsilon < \epsilon$$

for all $n > n_0$ and all $p \geqq 1$. Hence (cf. §3, Theorem 4) $\lim S_n$ exists.

---

[1] This means: if λ_n denotes the length of the longest segment in the nth decomposition, then $\lambda_n \to 0$.

Set this limiting value equal to J. We now obtain the theorem stated at the end of the preceding paragraph.

Theorem. *If $\epsilon > 0$ is given, and $\delta = \delta(\epsilon)$ is determined according to Lemma 3, then the relation*

$$|J_n - J| < \epsilon$$

holds for every Σ-sum J_n for which the lengths of all path segments are less than δ.

Proof: If, in the proof of Lemma 4, the number p in the inequality $|S_{n+p} - S_n| < \frac{1}{2}\epsilon$ is allowed to approach infinity, it follows first that

$$|S_n - J| \leqq \tfrac{1}{2}\epsilon \quad \text{for } n \geqq n_0.$$

Furthermore, by Lemma 3,

$$|S_n - J_n| < \tfrac{1}{2}\epsilon.$$

Hence

$$\begin{aligned}
|J_n - J| &= \left|(S_n - J) - (S_n - J_n)\right| \\
&\leqq |S_n - J| + |S_n - J_n| < \epsilon.
\end{aligned}$$

Thus the existence of the number J with the asserted properties, that is to say, the existence of the definite integral has been proved completely.

REMARKS. 1. Only the continuity of $f(z)$ along k was used in our proof, and not continuity in $\mathfrak{G}$. Hence, $f(z)$ need not even be defined except for k.

2. Our concept of integral includes the real integral (cf. §8, beginning) as a special case. To realize this, take k to be a segment of the real axis and $f(z)$ to be a function which is real-valued on k.

Exercise. Let $F(z)$ be a continuous function of z along k. Show that the limiting value

$$\lim_{n \to \infty} \left\{ \sum_{\nu=1}^{n} |z_\nu - z_{\nu-1}| F(\zeta_\nu) \right\} = \int_k F(z) \, |dz|,$$

understood in the same sense as before, always exists.

§10. Evaluation of Definite Integrals.

The problem of actually calculating the number J in given instances is of an entirely different nature. This is possible, in general, only under somewhat restrictive hypotheses.

Let us assume that the real functions

$$x = x(t), \quad y = y(t),$$

representing the coordinates of the point which describes the path as t traverses the interval $<\alpha, \beta>$, have *continuous derivatives* $x'(t)$ and $y'(t)$.

Then the path *is certainly rectifiable*. We decompose the path by dividing the parameter interval into n parts by means of the values

$$\alpha = t_0 < t_1 < t_2 < \cdots < t_n = \beta,$$

choosing intermediate parameter values $\tau_1, \tau_2, \ldots, \tau_n$, and setting

$$z_\nu = z(t_\nu) \quad \text{for} \quad \nu = 0, 1, 2, \ldots, n,$$

$$\zeta_\nu = z(\tau_\nu) \quad \text{for} \quad \nu = 1, 2, \ldots, n.$$

For brevity we set

$$u[x(t), y(t)] = \bar{u}(t), \quad v[x(t), y(t)] = \bar{v}(t).$$

Now we may write

$$\sum_{\nu=1}^{n} (z_\nu - z_{\nu-1}) f(\zeta_\nu)$$

$$= \sum_{\nu=1}^{n} [(x_\nu - x_{\nu-1}) + i(y_\nu - y_{\nu-1})] [\bar{u}(\tau_\nu) + i\bar{v}(\tau_\nu)].$$

Multiplying out the brackets we obtain four *real* Σ-sums. As we refine the subdivision, these sums tend to easily recognizable limits.

For example,

$$\sum_{\nu=1}^{n} (x_\nu - x_{\nu-1})\bar{u}(\tau_\nu) \quad \text{approaches} \quad \int_\alpha^\beta \bar{u}(t)x'(t)dt.$$

For, by the mean value theorem of the differential calculus,

$$x_\nu - x_{\nu-1} = x(t_\nu) - x(t_{\nu-1}) = (t_\nu - t_{\nu-1})x'(\tau_\nu'),$$

where τ_ν' denotes a value between $t_{\nu-1}$ and t_ν. Because of the assumed continuity of $x'(t)$ in $< \alpha, \beta >$ we may write

$$x'(\tau_\nu') = x'(\tau_\nu) + \epsilon_\nu,$$

where all ϵ_ν tend *uniformly* to zero as we refine the subdivision.[1]

Hence, the real Σ-sum in question is equal to

$$\sum_{\nu=1}^{n} (t_\nu - t_{\nu-1})x'(\tau_\nu) \cdot \bar{u}(\tau_\nu) + \sum_{\nu=1}^{n} (t_\nu - t_{\nu-1})\epsilon_\nu \cdot \bar{u}(\tau_\nu).$$

The first term in this expression is precisely that Σ-sum which tends to the *real* integral $\int_\alpha^\beta \bar{u}(t)x'(t)dt$. The second term, however, tends to zero, since, for given $\epsilon > 0$, it can be made smaller in absolute value than

$$\epsilon(\beta - \alpha) \cdot \bar{u}_0$$

by refining the subdivision. Here $\bar{u}_0$ denotes an upper bound of $|\bar{u}(t)|$ along k.

The other three Σ-sums may be treated in an analogous fashion.

According to this, the limit J (i.e., our definite

---

[1] This means that if $\epsilon > 0$ is given, there is a refinement of subdivision such that *all* $\epsilon_\nu < \epsilon$.

integral) which is approached by our complex Σ-sums has the value

$$(1) \quad J = \int_{z_0}^{Z} f(z)dz$$

$$= \int_{a}^{\beta} \bar{u}x'dt - \int_{a}^{\beta} \bar{v}y'dt + i \int_{a}^{\beta} \bar{u}y'dt + i \int_{a}^{\beta} \bar{v}x'dt.$$

We may write a condensed formula for J,

$$(2) \qquad J = \int_{a}^{\beta} (\bar{u} + i\bar{v})(x' + iy')dt,$$

which by this time will not be misunderstood; or still more briefly,

$$(3) \qquad J = \int_{a}^{\beta} f(z(t)) \cdot z'(t)dt;$$

or finally,

$$(4) \qquad J = \int_{a}^{\beta} f(z)dz = \int_{k} f(z)dz.$$

Here the limits with respect to t are to recall that z is a function of t, while the path alone is mentioned in the last form as the only essential. We see, in addition, that this investigation concerning the calculation of the value of the integral has given us a deeper insight into the meaning of the notation used for the definite integral.

Example 1.

$$f(z) = \frac{1}{z}; \quad k: z(t) = \cos t + i \sin t, \quad 0 \leqq t \leqq 2\pi.$$

The path is the unit circle described from $+1$ in the mathematically positive sense (counterclockwise) back to $+1$. Hence by (3),

$$J = \int_k \frac{dz}{z} = \int_0^{2\pi} \frac{1}{\cos t + i \sin t} (-\sin t + i \cos t)\, dt = i \int_0^{2\pi} dt = 2\pi i.$$

This result is used continually in the following sections.

Example 2.

$$f(z) = \Re(z) = x; \quad z_0 = 0, Z = 1 + i.$$

$\int_{z_0}^{Z} f(z)dz$ is to be evaluated along two distinct paths:

1. Path k_1: The straight line segment

$$z = (1 + i)\, t, \quad 0 \leqq t \leqq 1.$$

We have

$$J_1 = \int_0^1 t \cdot (1 + i)\, dt = (1 + i) \int_0^1 t\, dt = \tfrac{1}{2}(1 + i).$$

2. Path k_2: From 0 along a straight line to $+1$, and from there along a straight line to $1 + i$. By calculating both parts separately and adding the results we find that

$$J_2 = \tfrac{1}{2} + i.$$

Different values are thus obtained by using different paths. (Cf. §6, ex. 2.)

The following examples show that it is sometimes simplest to go back directly to the definition of the integral as the limit of a sum (§§ 8 and 9).

Example 3. Let $\mathfrak{G}$ be the entire plane; $f(z) = 1$; path: arbitrary.

We have

$$\begin{aligned}
J_n &= \sum_{\nu = 1}^{n} (z_\nu - z_{\nu-1}) \cdot 1 \\
&= (z_1 - z_0) + (z_2 - z_1) + \cdots + (Z - z_{n-1}) \\
&= Z - z_0.
\end{aligned}$$

Hence

$$\lim J_n = J = \int_{z_0}^{Z} dz = Z - z_0$$

along any path. If, in particular, k is a closed path, which we shall then denote by C,

$$\int_C dz = 0,$$

because $Z = z_0$.

Example 4. Let $\mathfrak{G}$ be the entire plane; $f(z) = z$; path: arbitrary.

We have

$$J_n = \sum_{\nu = 1}^{n} (z_\nu - z_{\nu - 1})\zeta_\nu,$$

where ζ_ν is an arbitrary point on that part of the path extending from $z_{\nu-1}$ to z_ν.

a) Take

$$\zeta_\nu = z_{\nu - 1}.$$

Then if the sum is denoted by J_n',

$$J_n' = (z_1 - z_0)z_0 + (z_2 - z_1)z_1 + \cdots + (Z - z_{n-1})z_{n-1}.$$

b) Take

$$\zeta_\nu = z_\nu.$$

If the sum is now denoted by J_n'', then

$$J_n'' = (z_1 - z_0)z_1 + (z_2 - z_1)z_2 + \cdots + (Z - z_{n-1})Z.$$

By addition it follows that

$$J_n' + J_n'' = Z^2 - z_0^2.$$

Consequently,

$$\lim (J_n' + J_n'') = 2J = Z^2 - z_0^2;$$

that is,

$$J = \int_{z_0}^{Z} zdz = \tfrac{1}{2} (Z^2 - z_0^2)$$

for an entirely arbitrary path. If k is a closed path C,

$$\int_C z\,dz = 0.$$

Example 5. $\int (z - z_0)^m dz$; path k: a circle with radius r about z_0 as center, described in the positive sense. k may be represented by

$$z = z_0 + r\,(\cos t + i \sin t), \quad 0 \le t \le 2\pi,$$

so that

$$J = \int_0^{2\pi} [r(\cos t + i \sin t)]^m \cdot r(-\sin t + i \cos t)\,dt$$

$$= ir^{m+1} \int_0^{2\pi} [\cos (m + 1) t + i \sin (m + 1)t]\,dt.$$

Now, as is well known,

$$\int_0^{2\pi} \cos \mu t\, dt = 0 \quad \text{and} \quad \int_0^{2\pi} \sin \mu t\, dt = 0$$

for every (positive or negative) integer μ distinct from zero, whereas for $\mu = 0$ the integrals are equal to

$$2\pi, \quad 0,$$

respectively. Hence, our integral

$$\int_k (z - z_0)^m\, dz = \begin{cases} 2\pi i & \text{for } m = -1 \quad (\text{cf. ex. 1}); \\ 0 & \text{for every other integral value of } m. \end{cases}$$

Exercises. 1. Evaluate the last integral also for the case that
a) k is a square whose center is z_0 and whose sides are parallel to the coordinate axes;
b) k is an ellipse whose center is z_0 and whose axes are parallel to the coordinate axes.

2. Evaluate $\int_{-i}^{+i} |z|\, dz$ by taking the path

a) rectilinearly, b) along the left half of the unit circle, c) along the right half of the unit circle.

§11. Elementary Integral Theorems

The following elementary theorems, in which the missing integrand should always read $f(z)dz$, follow almost immediately from the definition of the integral as the limit of a sum.

Theorem 1.

$$\int_{z_0}^{Z}{}_k + \int_{Z}^{Z'}{}_{k'} = \int_{z_0}^{Z'}{}_{(k+k')} \; ;$$

i.e., the sum of integrals taken along successive path segments is equal to the integral over the entire path. The notation $k + k'$ for the path of the integral on the right means that one is to proceed from z_0 to Z along k and then continue along k' to Z'.

Likewise

$$\int_{z_0}^{Z}{}_k = \int_{z_0}^{z'}{}_{k_1} + \int_{z'}^{Z}{}_{k_2},$$

if z' is chosen on k between z_0 and Z, thereby decomposing k into k_1 and k_2.

Theorem 2.

$$\int_{z_0}^{Z}{}_k = - \int_{Z}^{z_0}{}_k \; ;$$

i.e., if one integrates along the same path k, once in one direction and once in the opposite direction, then the two values obtained are the same except for sign. If one direction is denoted by $+ k$ and the other by $- k$, one can also write more briefly

$$\int_{-k} = - \int_{+k} \quad \text{or} \quad \int_{+k} + \int_{-k} = \int_{(+k)+(-k)} = 0.$$

*This can be stated briefly as follows: **If one integrates back and forth** over the same path, the value of the integral is zero.*

Theorem 3.

$$\int_k cf(z)dz = c\int_k f(z)dz;$$

i.e., a constant factor may be put before the integral sign.

Theorem 4.

$$\int_k [f_1(z) + f_2(z)]dz = \int_k f_1(z)dz + \int_k f_2(z)dz.$$

In words: the integral of a sum of two (or more, but still a finite number of) functions is equal to the sum of the integrals of the single terms.   Briefly, a sum (of a finite number of functions) may be integrated term by term.

Theorem 5.

$$\left| \int_k f(z)dz \right| \leqq Ml,$$

if M denotes a (positive) number which is not exceeded by $|f(z)|$ for any z on the path k, and l is the length of k.

The proof of this important formula follows immediately from the definition of the integral.   We have

$$J_n = \sum_{\nu=1}^{n} (z_\nu - z_{\nu-1})f(\zeta_\nu),$$

and hence

$$|J_n| \leqq \sum_{\nu=1}^{n} |z_\nu - z_{\nu-1}| \, |f(\zeta_\nu)| \leqq M \sum_{\nu=1}^{n} |z_\nu - z_{\nu-1}|.$$

The sum on the right, according to its meaning, represents the length of the segmental arc with the vertices $z_0, z_1, z_2, \ldots, Z$ inscribed in k, and hence is less than or equal to l for every n.

Consequently,

$$|J_n| \leqq Ml$$

for every n, and therefore also

$$|J| \leqq Ml, \qquad \text{Q. E. D.}$$

For instance, for the first example in §10, it follows without any computation that

$$\left| \int_k \frac{dz}{z} \right| \leqq 1 \cdot 2\pi = 2\pi,$$

since $|z| = 1$ for every point z of the unit circle k, and the length of the latter is 2π.

Exercise.   In connection with the exercise in §9, show that

$$\left| \int_k f(z)dz \right| \leqq \int_k |f(z)| \, |dz|.$$

CHAPTER 4

CAUCHY'S INTEGRAL THEOREM

§12. Formulation of the Theorem

According to the definition of the integral of a function of a complex variable, its value depends not only on the limits of integration z_0 and Z, (as is the case for a real integral), but also quite essentially on the path k which connects them (cf. §10, ex. 2). Now there is a theorem which states that, under hypotheses to be given immediately, such a dependence on the path does *not* exist if the function is not only continuous, as hitherto assumed, but also differentiable. This theorem, called **Cauchy's integral theorem** after its discoverer, is *fundamental* for the entire theory of functions.

The Fundamental Theorem of the Theory of Functions

First form. *Let the function $w = f(z)$ be regular in a simply connected region* $\mathfrak{G}$, *and let z_0 and Z be two (interior) points of* $\mathfrak{G}$. *Then the integral*

$$\int_{z_0}^{Z} f(z)dz$$

has the same value along every path of integration extending from z_0 to Z and lying entirely within $\mathfrak{G}$.

According to this, if k_1 and k_2 are two such paths which are distinct, we should have

$$\int_{k_1} f(z)dz = \int_{k_2} f(z)dz \quad \text{or} \quad \int_{k_1} - \int_{k_2} = 0.$$

47

By §11, 1 and 2 this could be interpreted as follows: the integral along a path beginning and terminating at z_0, that is to say, along a *closed* (although not necessarily simple) path C lying entirely within $\mathfrak{G}$ is zero. Thus, from the first form of the theorem follows the

Second form. *If $f(z)$ is regular in the simply connected region $\mathfrak{G}$, then*

$$\int_C f(z)dz = 0$$

if C denotes an arbitrary (not necessarily simple) closed path lying within $\mathfrak{G}$.

Conversely, the first form follows immediately from the second. For, let k_1 and k_2 be two arbitrary paths extending from z_0 to Z and lying within $\mathfrak{G}$. Then if $-k_2$ is joined to k_1, these together form a closed (although not always simple) path, so that we have

$$0 = \int_{k_1} - \int_{k_2} \quad \text{and hence} \quad \int_{k_2} = \int_{k_1}.$$

It is therefore sufficient to prove the fundamental theorem in the second form; and this will be done in the following paragraph in three steps: first, for the case that C is a triangle; then, that C is an arbitrary polygon; finally, that C is an arbitrary closed path.

In Examples 3 and 4 of §10 we already proved Cauchy's theorem for two special functions, namely, $f(z) = 1$ and $f(z) = z$; for it was shown that

$$\int_C dz = 0 \quad \text{and} \quad \int_C zdz = 0$$

for an arbitrary closed path C.

§13. Proof of the Fundamental Theorem

PART I. C is a triangle T lying with $\mathfrak{G}$.

Divide T into four congruent subtriangles[1] T^{I}, T^{II}, T^{III}, T^{IV} by means of segments parallel to the sides of T.   Then

$$\int_T = \int_{T^{\mathrm{I}}} + \int_{T^{\mathrm{II}}} + \int_{T^{\mathrm{III}}} + \int_{T^{\mathrm{IV}}}$$

if the paths of integration are all described in the mathematically positive sense.

For, as we integrate over the sides of the four subtriangles (cf. Fig. 1, in which the appropriate arrows are drawn inside each of the triangles) we integrate back and forth (cf. §11, 2) over the three auxiliary segments, so that their influence is automatically eliminated.

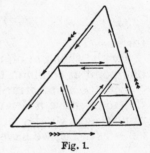

Fig. 1.

Of the four integrals on the right-hand side, there must be one, the path of which we denote by T_1, for which

$$\left| \int_T \right| \leqq 4 \left| \int_{T_1} \right| ,$$

since not every one of the four integrals can be less than one quarter of the whole.   The subtriangle T_1 can be treated in exactly the same way. T_1 contains at least one subtriangle T_2 for which

$$\left| \int_{T_1} \right| \leqq 4 \left| \int_{T_2} \right| ,$$

[1] The term "triangle" is used in two senses in this proof: the path, and the closed region determined by that path.   It will always be clear from the context, which of the two is meant at any particular time.

so that consequently

$$\left| \int_T \right| \leqq 4^2 \left| \int_{T_2} \right|.$$

Continuing in this manner, we obtain a sequence of similar triangles $T, T_1, T_2, \ldots, T_n, \ldots$ such that each lies inside the preceding one, is one quarter of the latter, and

$$\left| \int_T \right| \leqq 4^n \left| \int_{T_n} \right|$$

for $n = 1, 2, \ldots$.

By the theorem on nested sets, there is one and only one point z_0 common to all T_n; z_0 then also lies in $\mathfrak{G}$.

Let ϵ be an arbitrarily small positive number. Since $f(z)$ must have a derivative at z_0, $\delta > 0$ can be determined (see §6, II, first form) so that, for all z with $|z - z_0| < \delta$, we have

$$|f(z) - f(z_0) - (z - z_0)f'(z_0)| < \epsilon |z - z_0|,$$

or

$$f(z) = f(z_0) + (z - z_0)f'(z_0) + \eta \cdot (z - z_0)$$

with

$$|\eta| = |\eta(z)| < \epsilon.$$

Now choose n so large that T_n lies entirely within the neighborhood of z_0 characterized by $|z - z_0| < \delta$, so that $|z - z_0| < \delta$ for all z in the interior and on the boundary of T_n. Then

$$\int_{T_n} f(z)dz = \int_{T_r} f(z_0)dz - \int_{T_n} z_0 f'(z_0)dz$$

$$+ \int_{T_n} z f'(z_0)dz + \int_{T_n} \eta \cdot (z - z_0)dz.$$

Hence, by §11, 3 and the remark at the end of the preceding paragraph,

$$\int_{T_n} f(z)dz = 0 + 0 + 0 + \int_{T_n} \eta \cdot (z - z_0)dz,$$

and therefore by §11, 5,

$$\left| \int_{T_n} f(z)dz \right| < \epsilon \cdot \frac{s_n}{2} \cdot s_n = \frac{\epsilon}{2} \cdot s_n^2,$$

if s_n denotes the perimeter of T_n. This is true because $|z - z_0|$ is the distance between two points of one and the same triangle T_n and is therefore at most equal to $\frac{s_n}{2}$, the length of the path is s_n, and $|\eta| < \epsilon$.

Since

$$s_1 = \frac{s}{2}, \; s_2 = \frac{s_1}{2} = \frac{s}{2^2}, \ldots, s_n = \frac{s}{2^n}$$

if s denotes the perimeter of the given triangle T, we have finally

$$\left| \int_T f(z)dz \right| \leqq 4^n \left| \int_{T_n} f(z)dz \right| \leqq 4^n \cdot \frac{\epsilon}{2} \cdot \frac{s^2}{4^n} = \frac{\epsilon}{2} \cdot s^2.$$

The number on the right can be made arbitrarily small by the choice of ϵ, the value of the integral on the left is fixed. Therefore the latter must necessarily equal zero, Q. E. D.

PART II. The path C is an arbitrary closed polygon P which may intersect itself and which lies entirely within $\mathfrak{G}$.

First, if C is a quadrilateral Q which does not intersect itself, it can always be decomposed by means of a diagonal lying in its interior into two triangles T and T' which also lie within $\mathfrak{G}$, and we have again (cf. Fig. 2)

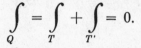

$$\int_Q = \int_T + \int_{T'} = 0.$$

Fig. 2.

By §4, Lemma 2, every arbitrary closed polygon P which does not intersect itself can likewise be decomposed into triangles by means of diagonals lying entirely in the interior of P. If one integrates over all these triangles separately, each of these integrals is equal to zero. If all of them are added together, the sum is equal to the integral taken along the boundary of the polygon P, since one integrates back and forth over every diagonal,[1] so that also

$$\int_P f(z)dz = 0.$$

Finally, by §4, Lemma 1, a closed polygon P which intersects itself can be decomposed into a finite number of closed polygons, each of which is simple and is described entirely in the positive or entirely in the negative sense; and possibly, in addition, a finite number of segments described twice, once in each direction. If one integrates over each part separately and adds, it is evident that again

$$\int_P f(z)dz = 0.$$

PART III. C is an arbitrary closed path lying within $\mathfrak{G}$.

Given $\epsilon > 0$, however small, we shall be able to find a suitable polygon P such that

[1] See Watson, *op. cit.*, p. 16, Theorem II.

$$\left| \int_C - \int_P \right| < \epsilon.$$

Then by II

$$\left| \int_C \right| < \epsilon; \quad \text{that is,} \quad \int_C = 0.$$

We recall that by definition

$$\int_C = \lim J_n = \lim \sum_{\nu=1}^{n} (z_\nu - z_{\nu-1}) f(\zeta_\nu), \quad \text{(with } z_0 = z_n).$$

After an arbitrary $\epsilon > 0$ has been given, choose the points of division z_ν so close together, and hence, n so large, that

1) $\left| \int_C - J_n \right|$ remains less than $\dfrac{\epsilon}{2}$, which is always

possible by the fundamental theorem of §9;

2) the lengths of all path segments are less than $\frac{1}{2}\rho$, where ρ is the number determined, according to Lemma 3 of §4, by C within $\mathfrak{G}$;

3) these lengths are also less than δ, if δ is a number such that

$$\left| f(z'') - f(z') \right| < \frac{\epsilon}{2l}, \qquad (l = \text{length of } C),$$

provided z' and z'' are any two points on C, or at a distance from C of at most $\frac{1}{2}\rho$, for which $|z'' - z'| < \delta$. Note that, in particular, if z denotes a point of the *chord* $z_{\nu-1} \ldots z_\nu$, we can set

$$f(z) = f(\zeta_\nu) + \eta_\nu, \quad \text{with} \quad |\eta_\nu| < \frac{\epsilon}{2l}.$$

The existence of δ follows from the theorem on uniform continuity.

If chords are now drawn from z_0 to z_1, from z_1 to z_2, ..., from z_{n-1} to $z_n = z_0$, a polygon P is formed which by 2) lies entirely with $\mathfrak{G}$. If one integrates along each side of P separately (hence, along a rectilinear path):

$$\int_P = \sum_{\nu=1}^{n} \int_{z_{\nu-1}}^{z_\nu} f(z)\,dz = \sum_{\nu=1}^{n} \int_{z_{\nu-1}}^{z_\nu} (f(\zeta_\nu) + \eta_\nu)\,dz$$

$$= \sum_{\nu=1}^{n} f(\zeta_\nu) \int_{z_{\nu-1}}^{z_\nu} dz + \sum_{\nu=1}^{n} \int_{z_{\nu-1}}^{z_\nu} \eta_\nu\,dz = J_n + \sum_{\nu=1}^{n} \int_{z_{\nu-1}}^{z_\nu} \eta_\nu\,dz,$$

and so

$$\left| \int_P - J_n \right| \leqq \sum_{\nu=1}^{n} \frac{\epsilon}{2l} \,|\, z_\nu - z_{\nu-1} \,| \leqq \frac{\epsilon}{2}.$$

Consequently,

$$\left| \int_C - \int_P \right| = \left| \left(\int_C - J_n \right) - \left(\int_P - J_n \right) \right|$$

$$\leqq \left| \int_C - J_n \right| + \left| \int_P - J_n \right| \leqq \tfrac{1}{2}\epsilon + \tfrac{1}{2}\epsilon = \epsilon.$$

Thus the polygon mentioned in the beginning of the proof has been obtained, and therefore

$$\int_C f(z)\,dz = 0, \qquad \text{Q. E. D.}$$

The third part of the proof says briefly this: since the integral over any polygon is always zero, and since an arbitrary path C can be approximated arbitrarily closely by an inscribed polygon, the integral taken along C cannot be different from zero.

§14. Simple Consequences and Extensions

Cauchy's integral theorem is the starting-point for almost all deeper investigations concerning analytic functions. All succeeding chapters will bear this out. Several simple consequences and extensions will be mentioned first.

1. If $\mathfrak{G}$ is an *arbitrary* region and $f(z)$ is regular in $\mathfrak{G}$, then

$$(1) \qquad\qquad \int_C f(z)dz = 0$$

for a closed path C if C can be imbedded in a simply connected subregion $\mathfrak{G}'$ of $\mathfrak{G}$; i.e., if there exists a simply connected subregion $\mathfrak{G}'$ of $\mathfrak{G}$ such that C lies within $\mathfrak{G}'$.

2. Since C is a continuum in the sense of §4, 3, the possibility stated in 1. always exists if the complementary set of $\mathfrak{G}$ lies entirely in the *outer region* determined in the plane by the continuum C. For a proof one has only to refer to Lemma 4 of §4 and chose the ϵ in it smaller than the distance of the path C from the set which is complementary to $\mathfrak{G}$. The interior of P then furnishes the simply connected subregion $\mathfrak{G}'$ of $\mathfrak{G}$ required in 1. In particular, equation (1) always holds when C is a simple closed path within $\mathfrak{G}$ whose interior belongs entirely to $\mathfrak{G}$.

3. We also have the somewhat deeper result that equation (1) is true for a simple closed path C if we know only that $f(z)$ is regular in the interior of C and at every point of the path itself.

A proof of this is given by E. Kamke, Math. Zeitschr., 35 (1932), pp. 539–543.

Even if $f(z)$ is only known to be regular in the interior of C and to assume a boundary value $f(z)$ at every point z of C (cf. §6), equation (1) holds again for these boundary values, which automatically form a continuous function along C (cf. §6, exercise 3). This extension of Cauchy's integral theorem is by no means self-evident; it was first proved by S. Pollard.[1]

4. Let C_1 and C_2 be two simple closed paths, C_2 lying entirely in the interior of C_1. Those points of the plane which lie both in the interior of C_1 and in the exterior of C_2 form a region which is called briefly the *annular region* determined by C_1 and C_2. If both paths lie within an arbitrary region $\mathfrak{G}$ in which $f(z)$ is regular, we have

Theorem 1.

$$\int_{C_1} f(z)dz = \int_{C_2} f(z)dz$$

if the annular region determined by C_1 and C_2 belongs entirely to $\mathfrak{G}$ and both paths are oriented in the same sense, whether the interior of C_2 belongs entirely to $\mathfrak{G}$ or not.

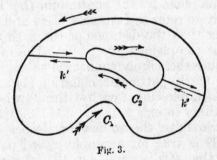

Fig. 3.

Proof: Connect (see Fig. 3) the paths C_1 and C_2 by means of two non-intersecting auxiliary paths k' and k''

---

[1] S. Pollard, Proc. London Math. Soc., 21 (1923), pp. 456–482. See also H. Heilbronn, Math. Zeitschr., 37 (1933), pp. 37–38; T. Estermann, *ibid.*, pp. 556–560, J. L. Walsh, Proc. Nat. Acad. Sci., 19 (1933), pp. 540–541. The best result of this kind, involving Lebesgue integration, was obtained by V. V. Golubev, Zap. Univ., otd. fiz.-mat. 29 (1916) (in Russian).

lying wholly within the annular region.[1] The latter is thereby decomposed into two simply connected subregions within which and on whose boundaries $f(z)$ is regular. By 2., the integrals over these boundaries equal zero, and hence their sum is also zero. However, by §11, 2, the integrals over the auxiliary paths are removed by adding, so that if C_1 and C_2 are both oriented in the mathematically positive sense, there remains

$$\int_{+C_1} + \int_{-C_2} = 0, \quad \text{that is,} \quad \int_{C_1} = \int_{C_2} \qquad \text{Q. E. D.}$$

Example.   By § 10, Example 5

$$\int_C \frac{dz}{z - z_0} = 2\pi i$$

if C is a circle about z_0 as center. According to the theorem just proved, this integral has the same value if C is any closed, simple, and positively oriented path whose interior contains z_0. Every one of these paths, taken as C_1, together with a sufficiently small circle with center z_0, taken as C_2, satisfies the hypothesis of Theorem 1. The analogue holds for every integral in §10, Example 5.

5. The following theorem is proved in an entirely similar manner.

Theorem 2. *Let C_0 be a simple closed path. Let each of the simple closed paths $C_1, C_2, \ldots, C_m$ lie wholly within the interior of C_0 but in the exterior of every other one of these paths (cf. Fig. 4, where $m = 3$). Then*

$$\int_{C_0} f(z)dz = \int_{C_1} f(z)dz + \int_{C_2} f(z)dz + \cdots + \int_{C_m} f(z)dz,$$

---

[1] It is easy to see that such auxiliary paths can always be drawn. For, consider two half-rays r_1, r_2 emanating from a point z_0 in the interior of C_2. If, beginning at z_0, the first point of intersection of r_1 with C_1 is denoted by B and the last point of intersection of the segment $z_0 \ldots B$ with C_2 is denoted by A, then $A \ldots B$ is such an auxiliary path; and one on r_2 is obtained by a similar argument.

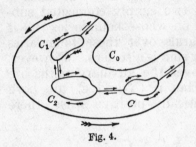

Fig. 4.

provided all the paths and the annular region between C_0 and the C_μ ($\mu = 1, 2, \ldots, m$) lie entirely within a region $\mathfrak{G}$ in which $f(z)$ is regular, and provided all the paths are oriented in the same sense.

The method of proof is suggested by the arrows in Fig. 4.

Example. By decomposing the integrand it is found that

$$\int_C \frac{2z - 1}{z^2 - z}\, dz = \int_{C_1} \frac{dz}{z} + \int_{C_2} \frac{dz}{z - 1} = 4\pi i,$$

if C encloses the points 0 and 1 whereas C_1, C_2 only enclose 0, 1, respectively.

6. We now can prove the existence of primitive functions of given regular functions. First we prove

Theorem 3. *If $f(z)$ is a continuous function in the simply connected region $\mathfrak{G}$, if z_0 is an arbitrary but fixed point of $\mathfrak{G}$, and if the integral[1]*

(1)

is independent of the path, provided the latter lies entirely within $\mathfrak{G}$, then the value of this integral is, in $\mathfrak{G}$, a regular function $F(z)$ of the upper limit of integration z. For this function, $F'(z) = f(z)$ for every z in $\mathfrak{G}$.

Proof: By hypothesis $F(z)$ is uniquely determined by the integral. As to the rest of the theorem, we must prove (see §6, II, first form) that

---

[1] The variable of integration in a definite integral may of course be designated quite arbitrarily. Here, as often in the following sections, it is called ζ, whereas z denotes an arbitrary point which is held fixed during the integration.

$$\left| \frac{F(z') - F(z)}{z' - z} - f(z) \right| < \epsilon$$

if z' lies sufficiently close to z. Since z is an interior point of $\mathfrak{G}$, a certain neighborhood of z lies entirely within $\mathfrak{G}$. Let z' be restricted to this neighborhood. By §11, 1

$$F(z') - F(z) = \int_z^{z'} f(\zeta)d\zeta,$$

where by hypothesis the path may be chosen arbitrarily. We take it to be rectilinear. Since the function $f(z)$ is continuous, we may set

$$f(\zeta) = f(z) + \eta,$$

where

$$|\eta| < \epsilon$$

for all ζ on the segment $z \ldots z'$ provided the neighborhood of the point z to which z' has been restricted is taken small enough. Then

$$F(z') - F(z) = (z' - z)f(z) + \int_z^{z'} \eta d\zeta,$$

whence by §11, 5

$$| F(z') - F(z) - (z' - z)f(z) | < \epsilon\, | z' - z |\,.$$

This implies the assertion stated in the beginning of the proof.

Example. According to this theorem, $\displaystyle\int_1^z \frac{d\zeta}{\zeta}$ is a regular function

in every simply connected region which contains the point $+1$ but not the point 0; e.g., the "right" half-plane (cf. §2, f).

Corollary. The hypotheses of Theorem 3 are certainly satisfied if $f(z)$ is regular in $\mathfrak{G}$. Hence, *every function which is regular in a simply connected region possesses a primitive function there.* This primitive function can be represented by the integral (1) of Theorem 3. It will be shown in §16, Theorem 4 that the independence of the integral (1) of the path, which is required in Theorem 3, *only* occurs when $f(z)$ is regular in $\mathfrak{G}$. Regular functions are thus the only ones to possess primitive functions.

7. We now have the following theorem which corresponds to the fundamental theorem of the differential and integral calculus.

Theorem 4. *If $f(z)$ is regular in the simply connected region $\mathfrak{G}$, and if $F(z)$ is a primitive function of $f(z)$ in $\mathfrak{G}$, then*

$$(2) \qquad \int_{z_0}^{z_1} f(z)dz = F(z_1) - F(z_0)$$

if the points z_0 and z_1 and the path of integration lie within $\mathfrak{G}$.

By Theorem 3, Corollary, and §7, Theorem 2, the integral (1) and the present primitive function $F(z)$ can differ by at most an additive constant:

$$\int_{z_0}^{z} f(z)dz = F(z) + c.$$

Letting $z = z_0$ it follows that $c = -F(z_0)$, and equation (2) is then obtained by setting $z = z_1$.

CHAPTER 5

CAUCHY'S INTEGRAL FORMULAS

§15. The Fundamental Formula

We shall now prove the most important consequence of Cauchy's theorem, namely, **Cauchy's integral formula.**

Theorem. *If $f(z)$ is regular in a region $\mathfrak{G}$, then the formula*

$$f(z) = \frac{1}{2\pi i} \int_C \frac{f(\zeta)}{\zeta - z} d\zeta$$

is valid for every simple, closed, positively oriented path C and every point z in its interior, provided C and its interior belong entirely to $\mathfrak{G}$.

This theorem states that if a function is known to be regular in a region $\mathfrak{G}$, and if its values are known along a closed simple path C in $\mathfrak{G}$ which does not enclose any point not belonging to $\mathfrak{G}$, then the values of the function in the interior of C are uniquely determined. It is evident from this interpretation that the theorem is quite remarkable. It shows that the values of a regular function are connected by a very strong bond so that the values along the boundary completely determine those in the interior of C. A similar situation is clearly impossible in the case of the most general and therefore the most arbitrary functions defined in §5. Later theorems will show that the bond mentioned is actually much stronger than that indicated by this theorem.

Proof: We have

$$\frac{1}{2\pi i} \int_C \frac{f(\zeta)}{\zeta - z} d\zeta = \frac{1}{2\pi i} \int_C \frac{f(z)}{\zeta - z} d\zeta + \frac{1}{2\pi i} \int_C \frac{f(\zeta) - f(z)}{\zeta - z} d\zeta.$$

By §11, Theorem 3 and §14, Theorem 1 (example), the first term $J_1 = f(z)$.[1] In the second, J_2, the path C may be replaced, according to §14, Theorem 1, by any other path (in the interior of C) enclosing the point z; e.g., by a small circle k with center z. Thus

$$J_2 = \frac{1}{2\pi i}\int_k \frac{f(\zeta) - f(z)}{\zeta - z}d\zeta.$$

Let the radius ρ of k be chosen so small that

$$|f(\zeta) - f(z)| < \epsilon$$

for every point ζ of k; this is certainly possible because of the continuity of $f(\zeta)$. Then by §11, 5,

$$|J_2| \leq \frac{1}{2\pi}\cdot\frac{\epsilon}{\rho}\cdot 2\pi\rho = \epsilon; \quad \text{that is,} \quad J_2 = 0.$$

Hence, we have

$$J_1 + J_2 = \frac{1}{2\pi i}\int_C \frac{f(\zeta)}{\zeta - z}d\zeta = f(z),$$

as was asserted.

§16. Integral Formulas for the Derivatives

If k is an arbitrary path and $\varphi(z)$ is a function defined and continuous along k, then the integral

$$(1) \qquad \frac{1}{2\pi i}\int_k \frac{\varphi(\zeta)}{\zeta - z}d\zeta$$

has a definite value for every z which does not lie on k, and hence, defines a single-valued function $f(z)$ for the

---

[1] Note that ζ here is the variable of integration and that z and $f(z)$ are to be regarded as constant.

points which do not belong to k.   We have the following theorem concerning this function.

Theorem 1.   *The function $f(z)$ defined by (1) is regular in every region $\mathfrak{G}$ which contains no point of k, and its derivative there is given by the formula*

$$(2) \qquad f'(z) = \frac{1}{2\pi i} \int_k \frac{\varphi(\zeta)}{(\zeta - z)^2} d\zeta.$$

Proof: For fixed z in $\mathfrak{G}$ it must be shown (cf. §6, II, third form) that

$$(3) \qquad \lim_{n \to \infty} \left\{ \frac{f(z_n) - f(z)}{z_n - z} - \frac{1}{2\pi i} \int_k \frac{\varphi(\zeta)}{(\zeta - z)^2} d\zeta \right\} = 0,$$

provided the z_n also lie in $\mathfrak{G}$ and tend to z.   Now by (1),

$$f(z) = \frac{1}{2\pi i} \int_k \frac{\varphi(\zeta)}{\zeta - z} d\zeta \quad \text{and} \quad f(z_n) = \frac{1}{2\pi i} \int_k \frac{\varphi(\zeta)}{\zeta - z_n} d\zeta.$$

Hence,

$$\frac{f(z_n) - f(z)}{z_n - z} = \frac{1}{2\pi i} \int_k \frac{\varphi(\zeta)}{z_n - z} \left[\frac{1}{\zeta - z_n} - \frac{1}{\zeta - z} \right] d\zeta$$

$$= \frac{1}{2\pi i} \int_k \frac{\varphi(\zeta)}{(\zeta - z)(\zeta - z_n)} d\zeta.$$

According to this, if the expression in the braces in assertion (3) is denoted by A_n,

$$A_n = \frac{1}{2\pi i} \int_k \varphi(\zeta) \left[\frac{1}{(\zeta - z)(\zeta - z_n)} - \frac{1}{(\zeta - z)^2} \right] d\zeta$$

$$= \frac{z_n - z}{2\pi i} \int_k \frac{\varphi(\zeta)}{(\zeta - z)^2 (\zeta - z_n)} d\zeta.$$

Let M be an upper bound of the values $| \varphi(\zeta) |$ along k. If the distance of the point z from k is denoted by d, and if n is chosen so large that $| z - z_n | < \frac{1}{2}d$, then it is evident by §11, 5 that

$$| A_n | < \frac{| z_n - z |}{2\pi} \cdot \frac{2M}{d^3} \cdot l$$

for such n. Hence

$$A_n \to 0, \qquad \text{Q. E. D.}$$

Formula (2) simply asserts that one may obtain the derivative of $f(z)$ by differentiating with respect to z under the integral sign in formula (1). One proves in an entirely similar manner that it is possible to repeat this any number of times.

Theorem 2. *The function $f(z)$ defined by* (1) *possesses derivatives in $\mathfrak{G}$ of every order. and these are given by the following formulas:*

$$(4) \qquad f''(z) = \frac{2!}{2\pi i} \int\limits_k \frac{\varphi(\zeta)}{(\zeta - z)^3} \, d\zeta,$$

and in general,

$$(5) \qquad f^{(n)}(z) = \frac{n!}{2\pi i} \int\limits_k \frac{\varphi(\zeta)}{(\zeta - z)^{n+1}} \, d\zeta$$

for $n = 1, 2, 3, \ldots$.[1]

We indicate the proof of (4). Using (2) we have

$$B_n = \frac{f'(z_n) - f'(z)}{z_n - z} - \frac{2!}{2\pi i} \int\limits_k \frac{\varphi(\zeta)}{(\zeta - z)^3} \, d\zeta$$

$$= \frac{1}{2\pi i} \int\limits_k \varphi(\zeta) \left[\frac{1}{z_n - z} \left(\frac{1}{(\zeta - z_n)^2} - \frac{1}{(\zeta - z)^2} \right) - \frac{2}{(\zeta - z)^3} \right] d\zeta.$$

[1] For $n = 0$, (5) also contains formula (1) if, as is customary, 0! is understood to have the value 1.

Then (4) is equivalent to the assertion: $B_n \to 0$. The expression in brackets in the integrand is equal to

$$(z_n - z) \frac{3\zeta - z - 2z_n}{(\zeta - z)^3 (\zeta - z_n)^2}.$$

Hence, if M_1 has a meaning similar to that of M above,

$$|B_n| < \frac{|z_n - z|}{2\pi} \cdot \frac{4M_1}{d^5} \cdot l; \text{ and consequently } B_n \to 0,$$

Q. E. D.

With the aid of this result, we are now in a position to derive an important property of regular functions. A single-valued function was said to be regular merely if it possesses a derivative. As is well known, in the case of functions of a real variable this implies nothing concerning the nature of this derivative; it need not even be continuous. For regular functions of a complex variable, however, we have the following very remarkable and fundamental theorem.

Theorem 3. *If a single-valued function $f(z)$ of a complex variable is defined in a region $\mathfrak{G}$ and has a first derivative there, then all higher derivatives exist (and are therefore continuous) in $\mathfrak{G}$.*

Proof: Let z be an arbitrary point in $\mathfrak{G}$, and let C be any simple closed path which contains z and only points of $\mathfrak{G}$ in its interior. Then by Theorem 1, since $f(z)$ is continuous along C,

$$\frac{1}{2\pi i} \int_C \frac{f(\zeta)}{\zeta - z} d\zeta$$

is a function which is regular and differentiable any number of times everywhere within C. By Cauchy's integral formula of §15, this function is the function $f(z)$ itself. Consequently it possesses derivatives of every order at z. Since z was chosen completely arbitrarily, the same conclusion is true for every point of $\mathfrak{G}$.

Corollary. *In addition to the fundamental formula, the formulas*

$$f^{(n)}(z) = \frac{n!}{2\pi i} \int\limits_C \frac{f(\zeta)}{(\zeta - z)^{n+1}} \, d\zeta,$$

$$(n = 1, 2, 3, \ldots)$$

are valid under the same hypotheses.

It follows from this fundamental result that the *converse of* Cauchy's integral theorem is true.

Theorem 4. *If $f(z)$ is continuous in the simply connected region $\mathfrak{G}$, and if*

$$\int\limits_C f(z)dz = 0$$

for every closed path C lying within $\mathfrak{G}$, then $f(z)$ is regular in $\mathfrak{G}$. (**Morera's Theorem.**)

Proof: Here, as in the deduction of the first form of the fundamental theorem from the second (§12), it follows that

$$\int\limits_{z_0}^{z} f(z)dz$$

is independent of the path and hence (cf. §14, Theorem 3) represents a function $F(z)$, regular in $\mathfrak{G}$, for which $F'(z) = f(z)$. By the preceding results, $F(z)$, as a regular function, has a second derivative in $\mathfrak{G}$; i.e., $f(z)$ has a first derivative in $\mathfrak{G}$. Hence, $f(z)$ is regular in $\mathfrak{G}$.

Exercise. Give a complete proof of formula (5) for $n = 3$, and in general, for arbitrary n.

SERIES AND THE EXPANSION OF ANALYTIC FUNCTIONS IN SERIES

CHAPTER 6

SERIES WITH VARIABLE TERMS

As already remarked in §3, we presume that the reader is familiar with the theory of infinite series with constant complex terms. We therefore turn immediately to a more general investigation concerning series with variable terms.

§17. Domain of Convergence

Let

$$f_0(z), f_1(z), \ldots, f_n(z), \ldots$$

be an infinite sequence of arbitrary functions (§5). Let there be certain points z which belong to the domains of definition of all of these functions. If z is such a point, then the series

$$f_0(z) + f_1(z) + f_2(z) + \cdots = \sum_{n=0}^{\infty} f_n(z)$$

may or may not converge. Denote by $\mathfrak{M}$ the set of all those points z for which all the terms are defined and for which the series is convergent. $\mathfrak{M}$ is called the *domain of convergence* of the given series.

The ordinary power series correspond to the special assumptions

$$f_n(z) = a_n z^n \quad \text{or} \quad f_n(z) = a_n (z - z_0)^n.$$

67

The first important property of such power series is that their domain of convergence $\mathfrak{M}$ is the interior of a certain circle about z_0 as center, the so-called *circle of convergence*, possibly with the inclusion of certain points of its circumference. We shall prove this fact by a method which will at the same time yield the radius of the circle of convergence.

Consider the sequence of non-negative real numbers

$$(1) \qquad |a_0|,\, |a_1|,\, |\sqrt{a_2}|,\, \ldots,\, |\sqrt[n]{a_n}|,\, \ldots$$

This sequence is certainly bounded on the left. We now prove the following

Theorem. *If the sequence* (1) *is also bounded on the right, and if μ is its upper limit (see §3), set*

a) $\qquad r = \dfrac{1}{\mu} \quad \text{if } \mu > 0,$

b) $\qquad r = \infty \quad \text{if } \mu = 0.$

If the sequence (1) *is not bounded on the right, set $r = 0$. Thus*

c) $\qquad r = 0 \quad \text{if } \mu = +\infty.$

Hence, if we use the proper interpretation, we have in all cases

$$r = \frac{1}{\mu} = \frac{1}{\lim \sqrt[n]{|a_n|}}.$$

The series $\Sigma a_n(z - z_0)^n$ is absolutely convergent for $|z - z_0| < r$, divergent for $|z - z_0| > r$. (**Cauchy-Hadamard theorem.**)

Proof: If we write z instead of $z - z_0$, it is evident that we may assume $z_0 = 0$.

a)   If $0 < \mu < +\infty$, then

$$\overline{\lim} \ \sqrt[n]{|a_n z^n|} = \mu |z| \begin{cases} < 1 \text{ when } |z| < \dfrac{1}{\mu}, \\[2mm] > 1 \text{ when } |z| > \dfrac{1}{\mu}. \end{cases}$$

By the radical test (see *Elem.*, §28), $\Sigma \, a_n z^n$ is absolutely convergent for the first z, divergent for the second z.

b) If $\mu = 0$, it must be shown that $\Sigma a_n z^n$ converges for every $z = z_1 (\neq 0)$.   Since now for nearly all n

$$\sqrt[n]{|a_n|} < \epsilon, \text{ e.g., } \sqrt[n]{|a_n|} < \frac{1}{2|z_1|}$$

and hence

$$\overline{\lim} \ \sqrt[n]{|a_n z_1^n|} \leqq \tfrac{1}{2},$$

the asserted convergence again follows immediately from the radical criterion.

c) Conversely, if $\Sigma a_n z^n$ is convergent for a $z = z_1 \neq 0$, then the sequence $\{a_n z_1^n\}$ is bounded.   Therefore the sequence $\{\sqrt[n]{|a_n|}\}$ is also bounded.   Hence, if $\mu = \infty$, our series can converge for no $z \neq 0$.

The theorem states nothing about the convergence or divergence of the series for the boundary points of the circle of convergence.   Indeed, the behavior of the series for such points varies from case to case: Σz^n is convergent for no boundary points; $\sum \dfrac{z^n}{n^2}$, for all boundary points; $\sum \dfrac{z^n}{n}$, for certain (but not all) boundary points.[1]

If the $f_n(z)$ are of a complicated nature, the determination of the exact domain of convergence is usually difficult.   In every case, however, the sum of a series $\Sigma f_n(z)$ is a definite number for every point of the

---

[1] For all three series, $r = 1$.

domain of convergence, and is therefore (cf. §5) a function $f(z)$ defined for all points of $\mathfrak{M}$. The infinite series is the prescribed rule by means of which a function is to be defined according to §5. One says: *the series represents the function $f(z)$ in $\mathfrak{M}$, or $f(z)$ can be expanded in the series there;* e.g., $\displaystyle\sum_{n=0}^{\infty} z^n$ represents the function $\dfrac{1}{1-z}$ in the unit circle, or $\dfrac{1}{1-z}$ can be expanded in that series there.

Since we have already recognized the regular functions as particularly valuable, the question arises: When does a series represent such a regular function? To be able to give a general answer to this question we need the concept of *uniform convergence* which will be developed in the following section.

Exercises. 1. Determine the radius of convergence of the power series $\displaystyle\sum_{n=1}^{\infty} a_n z^n$ if

$$\alpha)\ a_n = \frac{1}{n^n}; \quad \beta)\ a_n = n^{\log n}; \quad \gamma)\ a_n = \frac{n!}{n^n}.$$

2. Determine the domain of convergence of $\displaystyle\sum_{n=1}^{\infty} f_n(z)$ if

$$\alpha)\ f_n(z) = \frac{1}{n^z} = e^{-z \log n}, \qquad (\log n \geqq 0);$$

$$\beta)\ f_n(z) = \frac{z^n}{1 - z^n}.$$

That is to say, determine the domain of convergence of the series

$$\sum_{n=1}^{\infty} \frac{1}{n^z} \quad \text{and the series} \quad \sum_{n=1}^{\infty} \frac{z^n}{1 - z^n}.$$

§18. Uniform Convergence

Suppose the series $\Sigma f_n(z)$ has the domain of convergence $\mathfrak{M}$. This means that if z_1 is an arbitrary point of $\mathfrak{M}$ and $\epsilon > 0$ is given, we can determine a number $n_1 = n_1(\epsilon)$ such that

$$| f_{n+1}(z_1) + f_{n+2}(z_1) + \cdots + f_{n+p}(z_1) | < \epsilon$$

for all $n \geq n_1$ and all $p \geq 1$. If another point z_2 of $\mathfrak{M}$ is chosen, then, likewise, n_2 can be determined, etc. Thus, for a given ϵ, to every point z of $\mathfrak{M}$ there corresponds an integer $n_z = n_z(\epsilon)$ such that the absolute value of the sum of any finite number of consecutive terms after the n_zth term of the series for this value z is less than ϵ. Assume n_z to be taken as small as possible for given ϵ and z. The magnitude of n_z may be regarded as a measure of the rapidity of the convergence. If n_z is very large, the series converges slowly at the point z; if n_z is small, it converges rapidly.

Now suppose that there exists a number N which is greater than all the numbers n_z which correspond to the points z of $\mathfrak{M}$. Then, if $n \geq N$ and $p \geq 1$ are arbitrary,

$$| f_{n+1}(z) + f_{n+2}(z) + \cdots + f_{n+p}(z) | < \epsilon$$

for *every* point z in $\mathfrak{M}$; for, n now is also greater than every single n_z. Thus, the above-mentioned measure of the rapidity of convergence can be assigned for all points of $\mathfrak{M}$ in the same manner. We say briefly that the series **converges uniformly** in $\mathfrak{M}$. Hence, we have the following definition.

Definition. *The series $\Sigma f_n(z)$ converges uniformly in the domain*[1] *$\mathfrak{M}$ if, given $\epsilon > 0$, there exists a single positive integer $N = N(\epsilon)$ (depending only on ϵ and not on z) such that*

---

[1] Thus one can only speak of uniform convergence in *infinite* point sets $\mathfrak{M}$, never at single points; in particular, we consider uniform convergence in regions.

(1) $| f_{n+1}(z) + f_{n+2}(z) + \cdots + f_{n+p}(z) | < \epsilon$

for all $n \geqq N$, *all* $p \geqq 1$, *and all* z *in* $\mathfrak{M}$.

Since the series is assumed to converge at z, so that we may let p tend to infinity, it follows that if the series converges uniformly in $\mathfrak{M}$,

(2) $$\left| \sum_{\nu=n+1}^{\infty} f_\nu(z) \right| \leqq \epsilon$$

for all z in $\mathfrak{M}$ and all $n \geqq N$.

According to this, $\displaystyle\sum_{n=0}^{\infty} z^n$, for example, is *not* uniformly convergent in its domain of convergence (the unit circle); for, whatever n may be, $\displaystyle\sum_{\nu=n+1}^{\infty} z^\nu = \frac{z^{n+1}}{1-z}$ can actually be made arbitrarily large if z is only chosen on the segment $0 \cdots + 1$ near enough to $+1$. This example, at the same time, proves that a power series need *not* converge uniformly in its entire circle of convergence. On the other hand, we have the following theorem.

Theorem 1. *A power series converges uniformly in every circle which is smaller than and concentric to its circle of convergence. Thus, the uniformity of the convergence can only be disturbed near the circumference.*

Proof: Let $\Sigma a_n(z - z_0)^n$ have the radius of convergence $r > 0$. Let $0 < \rho < r$, and let z be an arbitrary point for which $| z - z_0 | \leqq \rho$. Then

$$\left| \sum_{\nu=n+1}^{n+p} a_\nu(z - z_0)^\nu \right| \leqq \sum_{\nu=n+1}^{n+p} | a_\nu | \rho^\nu$$

for all these z. But $\Sigma | a_n | \rho^n$ is convergent, since the point $z = z_0 + \rho$ lies in the interior of the circle of convergence. Hence, given $\epsilon > 0$, we can assign a number N such that

$$| a_{n+1} | \rho^{n+1} + \cdots + | a_{n+p} | \rho^{n+p} < \epsilon$$

for all $n \geqq N$ and all $p \geqq 1$.   Then likewise

$$| a_{n+1}(z - z_0)^{n+1} + \cdots + a_{n+p}(z - z_0)^{n+p} | < \epsilon$$

for all $| z - z_0 | \leqq \rho$, all $n \geqq N$, and all $p \geqq 1$,

Q. E. D.

There is the following general criterion for uniform convergence, which is called the **Weierstrass M-test**.

Theorem 2. *If the positive numbers M_0, M_1, ... M_n, . . . are such that*

$$| f_n(z) | \leqq M_n, \quad (n = 0, 1, 2, \ldots),$$

for all z of a subdomain $\mathfrak{M}'$ of the domain of convergence of the series $\Sigma f_n(z)$, and such that

$$\sum_{n=0}^{\infty} M_n$$

converges, then $\Sigma f_n(z)$ is uniformly convergent in $\mathfrak{M}'$.

The proof is entirely analogous to that of the special case just considered

Exercises.   1. Investigate the series given in §17, Exercise 2 as to uniformity of convergence.

2. Prove that the power series $\displaystyle\sum_{n=1}^{\infty} \frac{z^n}{n^2}$ converges uniformly in its *entire* circle of convergence.

§19. Uniformly Convergent Series of Analytic Functions

We now make the further assumption that all of the functions $f_n(z)$ are analytic.   We shall then show that the function represented by the series is also analytic. More precisely, let $f_0(z)$, $f_1(z)$, . . . be an infinite sequence of functions, all of which are regular in the same simply connected region $\mathfrak{G}$, and let the series

$\Sigma f_n(z)$ be *uniformly* convergent in every closed subregion $\mathfrak{G}'$ of $\mathfrak{G}$.[1] Then the following three theorems hold.

Theorem 1. *The series $\Sigma f_n(z)$ represents a function $F(z)$ which is continuous in $\mathfrak{G}$.*

Theorem 2. *Every series obtained by integrating term by term along a path k in $\mathfrak{G}$ converges and furnishes the integral of $F(z)$; in symbols:*

$$\sum_{n=0}^{\infty} \int_k f_n(z)dz \quad converges \ and \ is \ equal \ to \int_k F(z)dz.$$

Theorem 3. *$F(z)$ is a regular function in $\mathfrak{G}$, and every series obtained by differentiating p times term by term converges everywhere in $\mathfrak{G}$, in fact, uniformly in every closed subregion $\mathfrak{G}'$ of $\mathfrak{G}$, and furnishes the corresponding derivative of $F(z)$ there. In symbols, for fixed $p = 0$, $1, 2, \ldots,$* $\sum_{n=0}^{\infty} f_n^{(p)}(z)$ *converges in $\mathfrak{G}$ and is equal to $F^{(p)}(z)$.*

Proofs:

1. Given z_0 in $\mathfrak{G}$ and $\epsilon > 0$, it suffices to show that

$$| F(z) - F(z_0) | = | \Sigma f_n(z) - \Sigma f_n(z_0) | < 3\epsilon$$

for all z of $\mathfrak{G}$ which lie sufficiently close to z_0. To this end, we first choose a circle $\mathfrak{G}'$ which (inclusive of its boundary) lies within $\mathfrak{G}$ and has z_0 for its center. Set

$$\sum_{n=0}^{N} f_n(z) = A(z) \quad \text{and} \quad \sum_{n=N+1}^{\infty} f_n(z) = R(z).$$

Then, according to §18, there exists an N such that

$$| R(z) | \leqq \epsilon$$

for all z in $\mathfrak{G}'$. Let z be restricted to such a small neighborhood of z_0 within $\mathfrak{G}'$ that

---

[1] I.e., in every closed region $\mathfrak{G}'$ which, inclusive of its boundary points, belongs to the interior of the region $\mathfrak{G}$.

$$| A(z) - A(z_0) | < \epsilon$$

for all z there. Such a neighborhood can certainly be determined, since $A(z)$ is the sum of a finite number of continuous functions, and therefore continuous. We have

$$| F(z) - F(z_0) | \leqq | A(z) - A(z_0) | + | R(z) | + | R(z_0) |$$
$$< \epsilon + \epsilon + \epsilon = 3\epsilon,$$

Q. E. D.

2. Since $F(z)$ has been shown to be a continuous function, the integral of $F(z)$ appearing in the second heorem exists in any case. Indeed, by §11 Theorem 4

$$\int_k F(z)dz = \int_k A(z)dz + \int_k R(z)dz.$$

By the same theorem

$$\int_k A(z)dz = \int_k f_0(z)dz + \int_k f_1(z)dz + \cdots + \int_k f_N(z)dz.$$

Hence

$$\left| \int_k F(z)dz - \sum_{n=0}^{N} \int_k f_n(z)dz \right| = \left| \int_k R(z)dz \right| \leqq \epsilon \cdot l,$$

if l denotes the length of the path k. Since $\epsilon \cdot l$ can be made arbitrarily small by suitable choice of ϵ, this means that

$$\sum_{n=0}^{\infty} \int_k f_n(z)dz \quad \text{converges and is equal to} \int_k F(z)dz.$$

It now becomes evident that uniform convergence of $\Sigma f_n(z)$ along the path k is sufficient, and we can state the following theorem (an extension of §11, 4).

Theorem 4.   *An infinite series of continuous functions may be integrated term by term, provided that the series is uniformly convergent along the path of integration.*

3. If C is an arbitrary closed path lying within $\mathfrak{G}$, then $\Sigma f_n(z)$ is uniformly convergent along C. Hence by 2,

$$\int_C F(z)dz = \int_C \left(\sum f_n(z)\right)dz = \sum \int_C f_n(z)dz,$$

which equals zero since each term is equal to zero by virtue of Cauchy's integral theorem. Since C was chosen arbitrarily within $\mathfrak{G}$, $F(z)$ is regular in $\mathfrak{G}$ by Morera's theorem (§16, Theorem 4).

Now let $\mathfrak{G}'$ be any closed subregion of $\mathfrak{G}$. Then, according to §4, Lemma 3, C can be chosen so that it encloses $\mathfrak{G}'$ without having a point in common with it, so that consequently the distance ρ of C from $\mathfrak{G}'$ is still positive. For the pth derivative at the point z of $\mathfrak{G}$ (this derivative certainly exists), we obtain, for the same reasons as above,

$$F^p(z) = \frac{p!}{2\pi i} \int_C \frac{F(\zeta)}{(\zeta - z)^{p+1}} d\zeta$$

$$= \sum_{n=0}^{\infty} \frac{p!}{2\pi i} \int_C \frac{f_n(\zeta)}{(\zeta - z)^{p+1}} d\zeta = \sum_{n=0}^{\infty} f_n^{(p)}(z),$$

which proves the second part of the theorem. That this series actually converges uniformly in $\mathfrak{G}$ for fixed p follows from the simple inequality

$$\left| \sum_{\nu=n+1}^{n+r} f_\nu^{(p)}(z) \right| = \left| \frac{p!}{2\pi i} \int_C \frac{\sum_{\nu=n+1}^{n+r} f_\nu(\zeta)}{(\zeta - z)^{p+1}} d\zeta \right| \leq \frac{p!}{2\pi} \cdot l \frac{\epsilon}{\rho^{p+1}}.$$

(Also see the following Exercise 2 in this respect.)

Application to power series.

1. Let $f_n(z) = a_n(z - z_0)^n$, $(n = 0, 1, 2, \ldots)$, so that $\Sigma f_n(z)$ becomes the power series $\Sigma a_n(z - z_0)^n$. Let the radius of convergence r (§17) be greater than zero and let ρ be chosen between 0 and r, $(0 < \rho < r)$. Then the circle $| z - z_0 | < r$ can be taken as the region $\mathfrak{G}$, and, according to §18, Theorem 1, the circle with radius ρ and center z_0 can be taken as the subregion $\mathfrak{G}'$. Hence, we have

Theorem 5. *A power series $\Sigma a_n(z - z_0)^n$, within its circle of convergence, represents a regular function $f(z)$ whose derivatives are obtained by differentiating the power series term by term, and these derived power series have the same radius of convergence as the given series:*

$$f^{(p)}(z) = \sum_{n=0}^{\infty} n(n - 1) \ldots (n - p + 1)a_n(z - z_0)^{n-p}$$

$$= \sum_{n=0}^{\infty} (n + 1)(n + 2) \ldots (n + p)a_{n+p}(z - z_0)^n$$

$$= p! \sum_{n=0}^{\infty} \binom{n + p}{p} a_{n+p}(z - z_0)^n$$

converges for $| z - z_0 | < r.$

In particular,

$$f^p(z_0) = p! \, a_p, \quad a_p = \frac{1}{p!} f^p(z_0) = \frac{1}{2\pi i} \int_C \frac{f(\zeta)}{(\zeta - z_0)^{p+1}} \, d\zeta,$$

if C denotes the circumference $| z - z_0 | = \rho$. From the last formula we get, writing n instead of p, the following useful inequality known as **Cauchy's inequality:**

$$| a_n | \leqq \frac{1}{2\pi} \cdot 2\pi\rho \cdot \frac{M}{\rho^{n+1}} = \frac{M}{\rho^n},$$

if M is the maximum of $| f(z) |$ on $| z - z_0 | = \rho.$

Exercises. 1. Determine whether the series given in §17, Exercise 2 represent (within their regions of convergence) analytic functions.

2. In connection with the exercises of §§9 and 11, show that if, in addition to the series $\Sigma f_n(z)$, the series $\Sigma \mid f_n(z) \mid$ also converges uniformly in every $\mathfrak{G}'$, then Theorem 3 can be sharpened to the effect that the series $\Sigma \mid f^{(p)}(z) \mid$, for fixed p, also converge uniformly in $\mathfrak{G}'$.

CHAPTER 7

THE EXPANSION OF ANALYTIC FUNCTIONS IN POWER SERIES

The theorems of the preceding chapter show that the property of representing regular functions, possessed by power series in their regions of convergence, is shared by much more general series, namely, all *uniformly* convergent series whose terms are themselves regular functions. The great importance of power series for the study of analytic functions therefore cannot be based on this property. It rests, rather, on its converse: every regular function can be represented by a power series. Thus, the totality of all possible power series also furnishes the totality of all conceivable regular functions.

§ 20. Expansion and Identity Theorems for Power Series

Theorem 1. *Let $f(z)$ be a function regular in a certain region $\mathfrak{G}$ and let z_0 be an interior point of $\mathfrak{G}$. Then there is always one and only one power series of the form*

$$\sum_{n=0}^{\infty} a_n(z - z_0)^n$$

which converges for a certain neighborhood of z_0 and represents the function $f(z)$ in that neighborhood. Moreover,

$$a_n = \frac{1}{n!} f^{(n)}(z_0).$$

The series converges at least in the largest circle about the center z_0, which encloses only points of $\mathfrak{G}$; and the exact radius of convergence of the series is the largest circle (let its radius be r) about z_0 as center in which $f(z)$ is every-

where defined or definable as a differentiable function.
(**Expansion theorem; Taylor expansion.**)

Proof: Let z be an arbitrary interior point of the circle with radius r and center z_0. Then we must first show that for the given values of a_n,

$$\sum_{n=0}^{\infty} a_n(z - z_0)^n \text{ converges and is equal to } f(z).$$

Since $|z - z_0| = \rho < r$, we can choose ρ_1 so that $\rho < \rho_1 < r$. Let ζ be an arbitrary point of the circumference k_1 of the circle with radius ρ_1 and center z_0. Then

$$\frac{1}{\zeta - z} = \frac{1}{(\zeta - z_0) - (z - z_0)} = \frac{1}{\zeta - z_0} \cdot \frac{1}{1 - \dfrac{z - z_0}{\zeta - z_0}}$$

$$= \sum_{n=0}^{\infty} \frac{(z - z_0)^n}{(\zeta - z_0)^{n+1}}.$$

This particular geometric series is uniformly convergent with respect to ζ along k_1 (by §18, Theorem 2), since

$$\left|\frac{z - z_0}{\zeta - z_0}\right| = \frac{\rho}{\rho_1} < 1.$$

The same is true for the series

$$\frac{f(\zeta)}{\zeta - z} = \sum_{n=0}^{\infty} \frac{f(\zeta)}{(\zeta - z_0)^{n+1}} (z - z_0)^n.$$

Hence, if we integrate both sides along the path k_1, the integration on the right-hand side may be carried out term by term and we are certain, by §19, Theorem 2, that the resulting series is convergent. Dividing by $2\pi i$ we have therefore

$$\frac{1}{2\pi i} \int_{k_1} \frac{f(\zeta)}{\zeta - z} d\zeta = \sum_{n=0}^{\infty} \frac{1}{2\pi i} \int_{k_1} \frac{f(\zeta)}{(\zeta - z_0)^{n+1}} (z - z_0)^n d\zeta,$$

and hence by §§15 and 16

$$f(z) = \sum_{n=0}^{\infty} \frac{1}{n!} f^{(n)}(z_0) \cdot (z - z_0)^n = \sum_{n=0}^{\infty} a_n(z - z_0)^n,$$

Q. E. D.

That the expansion obtained is the only possible one follows immediately from the following **identity theorem for power series.**

Theorem 2. *If both power series*

$$\sum_{n=0}^{\infty} a_n(z - z_0)^n \quad and \quad \sum_{n=0}^{\infty} b_n(z - z_0)^n$$

have a positive radius of convergence, and if their sums coincide for all points of a neighborhood of z_0, or only for an infinite number of such points (distinct from one another and from z_0) with the limit point z_0, then they are identical.

Proof: First, for $z = z_0$ it follows that $a_0 = b_0$. Assume that the first m coefficients of both expansions have been proved to be the same, respectively. Then we have

$$a_{m+1} + a_{m+2}(z - z_0) + \cdots = b_{m+1} + b_{m+2}(z - z_0) + \cdots$$

for all of those infinitely many points. If in this equality we let z approach the limit point z_0 by means of those points, since the power series represent continuous functions it follows from §6, I, third form, that

$$b_{m+1} = a_{m+1}.$$

Hence, both expansions are identical.

Example. It is shown in §14, 6 that $f(z) = \displaystyle\int_{1} \frac{d\zeta}{\zeta}$ is a regular

function of z, if z and the (otherwise arbitrary) path of integration are confined to the interior of the right half-plane. $f(z)$

must therefore admit of a power series expansion, say for a neighborhood of $z_0 = +1$, for which r is at least unity. Since

$$f'(z) = \frac{1}{z}, f''(z) = -\frac{1}{z^2}, \cdots, f^n(z) = (-1)^{n-1} \frac{(n-1)!}{z^n}, \cdots,$$

we have, for $z = z_0 = 1$:

$$a_0 = 0, \; a_1 = 1, \; a_2 = -\frac{1}{2}, \cdots, a_n = \frac{(-1)^{n-1}}{n}, \cdots,$$

so that

$$f(z) = (z-1) - \frac{1}{2}(z-1)^2 + \cdots + \frac{(-1)^{n-1}}{n}(z-1)^n + \cdots$$

This is the only possible expansion. We see that $r = 1$.

An expansion of a function of a *real* variable in a power series does not always exist even if the function has derivatives of every order. Here, however, everything was deduced merely from the *existence* of the *first* derivative.

The expansion obtained converges, as already emphasized, in the largest circle K, with center z_0, in whose interior $f(z)$ is still everywhere defined or *definable* as a differentiable function. The latter means the following. Let $f(z)$ be defined in $\mathfrak{G}$, and let K be a circle, with center z_0, which may include points not contained in $\mathfrak{G}$. It may be possible to define the function $f(z)$ at the points of K which are not points of $\mathfrak{G}$ in such a manner that the resulting function is differentiable in the whole circle K. Then our power series converges at least in K. Furthermore, let K_0 be the largest circle, with center z_0, possessing the above property. Then K_0 is the exact circle of convergence of the power series. Note that two extreme cases are possible: K_0 may lie within the region $\mathfrak{G}$ where $f(z)$ was originally defined, or K_0 may be the entire plane. If it is impossible to include some point in a circle of convergence of a power series representing the function $f(z)$, this point is called a *singular point* of the function.

These remarkable matters will be treated in detail in the next chapter and in Section IV. At this time, only two special results will be deduced from our theorems.

The following theorem, called **Weierstrass's double-series theorem**, is often used to advantage in obtaining the power-series expansion of a given function.

Theorem 3.   *Let all of the functions*

$$f_n(z) = \sum_{k=0}^{\infty} a_k{}^{(n)}(z - z_0)^k,$$

$n = 0, 1, 2, \ldots$, *be regular at least for* $|z - z_0| < r$, *and let*

$$F(z) = \sum_{n=0}^{\infty} f_n(z)$$

$$= [a_0^{(0)} + a_1^{(0)}(z - z_0) + \cdots + a_k^{(0)}(z - z_0)^k + \cdots]$$
$$+ [a_0^{(1)} + a_1^{(1)}(z - z_0) + \cdots + a_k^{(1)}(z - z_0)^k + \cdots]$$
$$+ \ldots\ldots\ldots\ldots\ldots\ldots\ldots\ldots\ldots\ldots\ldots\ldots\ldots$$
$$+ [a_0^{(n)} + a_1^{(n)}(z - z_0) + \cdots + a_k^{(n)}(z - z_0)^k + \cdots]$$
$$+ \ldots\ldots\ldots\ldots\ldots\ldots\ldots\ldots\ldots\ldots\ldots\ldots\ldots$$

be uniformly convergent for $|z - z_0| \leqq \rho < r$ *for every* $\rho < r$. *Then the coefficients in any column form a convergent series; and if we set*

$$a_k^{(0)} + a_k^{(1)} + \cdots + a_k^{(n)} + \cdots = \sum_{n=0}^{\infty} a_k^{(n)} = A_k$$

for $k = 0, 1, 2, \ldots$, *then*

$$\sum_{k=0}^{\infty} A_k(z - z_0)^k$$

is the power series for $F(z)$; *it converges at least for* $|z - z_0| < r$.[1]

Proof: According to §19, Theorem 3, $F(z)$ is regular for $|z - z_0| < r$, and hence, by the expansion theorem,

---

[1] I.e., under the above hypotheses, the infinitely many power series may be added term by term.

can be developed in a power series there. Its kth coefficient is equal to

$$\frac{1}{k!} F^{(k)}(z_0) = \sum_{n=0}^{\infty} \frac{1}{k!} f_n^{(k)}(z_0) = \sum_{n=0}^{\infty} a_k^{(n)} = A_k,$$

which already completes the proof.

We prove finally the remarkable and important

Theorem 4. *An analytic function $f(z)$ cannot have a maximum modulus[1] at a point z_0 of a region of regularity, unless $f(z)$ has the same value $f(z_0)$ everywhere in that region.*

Proof: In a neighborhood of z_0 we have

$$f(z) = a_0 + a_1(z - z_0) + a_2(z - z_0)^2 + \cdots,$$
$$\text{(with } r > 0).$$

Let at least one of the coefficients following $a_0 = f(z_0)$ be different from zero, and let a_m, $(m \geqq 1)$, be the first such coefficient. Set

$$a_0 = Ae^{ia}, \ a_m = A'e^{ia'}, \ (A' > 0), \ z - z_0 = \rho e^{i\varphi},$$
$$(0 < \rho < r),$$

so that

$$f(z) = Ae^{ia} + A'e^{ia'}\rho^m e^{im\varphi} + a_{m+1}(z - z_0)^{m+1} + \cdots.$$

We now choose φ so that $\alpha' + m\varphi = \alpha$.[2] Then

$$f(z) = (A + A'\rho^m)e^{ia} + a_{m+1}(z - z_0)^{m+1} + \cdots,$$
$$|f(z)| \geqq A + A'\rho^m - (|a_{m+1}|\rho^{m+1} + \cdots)$$
$$\geqq A + \rho^m[A' - (|a_{m+1}|\rho + \cdots)].$$

Because of the continuity of the power series in the parentheses, we can take ρ here to be such a small number ρ_0 that $(|a_{m+1}|\rho_0 + \cdots) < \frac{1}{2}A'$. Then

$$|f(z)| > A + \frac{1}{2}A'\rho^m > |f(z_0)|$$

---

[1] I.e. a value which in absolute value is greater than or equal to all values of $f(z)$ in a neighborhood of z_0.

[2] I.e., we select a particular one of the radii emanating from z_0.

for all ρ with $0 < \rho < \rho_0$. That is, for all points z lying sufficiently close to z_0 on a certain radius emanating from z_0 we have $| f(z) | > | f(z_0) |$.

The following theorem, which is called the **principle of the maximum modulus,** is only a rewording of this result.

Theorem 5. *The maximum modulus of a function which is regular in a closed region always lies on the boundary of that region.*

Exercise. Expand the series given in §17, Exercise 2 in power series with the center $z_0 = + 2$ (for the first) and $z_0 = 0$ (for the second).

§21. The Identity Theorem for Analytic Functions

Cauchy's theorem and Taylor's expansion of a regular function (obtained by means of Cauchy's theorem) lead to most important results. These results will divulge the true nature of regular analytic functions. We start with a few preliminary remarks in this direction.

In §5 the most general concept of a function was given. This concept includes such arbitrary functions that it is impossible to infer anything from the behavior of such a function in one part of its region of definition $\mathfrak{M}$ as to its behavior in another part of this region. For instance, let $\mathfrak{M}$ be the entire plane and let $f(z) = 3i$ for $| z | \leqq 1$. Nothing can be said about the values of $f(z)$ for $| z | > 1$. Indeed, values may be assigned there according to a completely new defining rule (cf. the example on p. 22). The situation is different if $f(z)$ is required to be continuous. Then in the last example $f(z)$ must be close to $3i$ for points z near the unit circle. Thus, the condition of continuity restricts the function. It introduces a certain connection between its values, some kind of an intrinsic order. This connection permits us to say something about the values of the function in one part of the z plane if we know its values in another *adjacent* part. It is clear that this inner bond becomes stronger as we restrict the function to more special classes. An example from the theory of functions of a real variable x will clarify this matter.

Suppose we restrict our investigation to the class of entire rational functions (polynomials) of the third degree (i.e. to curves of the third degree):

$$y = a_0 + a_1x + a_2x^2 + a_3x^3, \quad (a_\nu, x, y \text{ real}).$$

Such a function is already completely determined by very few conditions (requirements). If we know, for example, that the curve passes through four specific distinct points (i.e., if we know the values of the function for four distinct values of x), the function is fully defined, no matter how close to one another the four points may lie. The behavior of the curve, with all its regular and singular properties, in the whole xy-plane can thus be inferred from the behavior of the function in an arbitrary small interval. The class of polynomials of the third degree exhibits a very strong inner bond by means of which the values of the function are linked together.

Since natural phenomena themselves possess an intrinsic regularity, it is clear that, above all, those functions which possess such an inner structure will appear in applications in the natural sciences.

Now, it is exceedingly remarkable that by means of the *single* requirement of differentiability, that is, *the requirement of regularity*, a class of functions having the following properties is selected from the totality of the most general functions of a complex variable. On the one hand, this class is still very general and includes almost all functions arising in applications. On the other hand, a function belonging to this class possesses such a strong inner bond, that from its behavior in a region, however small, of the z-plane one can deduce its behavior in the entire remaining part of the plane. To anticipate the most important result, we shall show that an analytic function, with all its regular and singular properties, is fully determined if the values of the function are known along any small arc. In other words, two analytic functions which coincide along such an arc are completely identical.

A first theorem in this direction is Cauchy's formula (cf. the discussion on p. 61) which enables us to deduce the values of the function in the interior of a simple closed path C from the values along the boundary. A second result of this kind is the statement made in connection with the expansion theorem as to the magnitude of the true circle of convergence of a power series. Indeed, here we have already taken into consideration points of the plane which do not even belong to the original region of definition of the function.

On the basis of the expansion theorem we are now in a position to derive a result which leads to the theorem stated and even beyond. Because of its great importance for the development of the theory of functions, it is the most fundamental result after Cauchy's integral theorem.

The identity theorem for analytic functions. *If two functions are regular in a region $\mathfrak{G}$, and if they coincide in a neighborhood, however small, of a point z_0 of $\mathfrak{G}$, or only along a path segment, however small, terminating in z_0, or also only for an infinite number of distinct points with the limit point z_0, then the two functions are equal everywhere in $\mathfrak{G}$.*

Proof: Denote the two functions by $f_1(z)$ and $f_2(z)$ and let K_0 be the largest circle with center z_0 which lies entirely within $\mathfrak{G}$. By virtue of the expansion theorem, both functions may be developed in power series which converge at least in K_0. On the basis of our hypotheses, the identity theorem for power series implies the identity of the two expansions. Therefore $f_1(z) = f_2(z)$ everywhere in K_0.

Now let ζ be an arbitrary point of $\mathfrak{G}$; we must show that we also have $f_1(\zeta) = f_2(\zeta)$. To this end, connect (see Fig. 5) z_0 and ζ by means of a path k lying entirely within $\mathfrak{G}$. Let ρ be the positive number whose existence is proved in §4, Lemma 3. Divide the path k in any manner (by means of points of division z_0, z_1, z_2, ..., z_{m-1}, $z_m = \zeta$) into subpaths whose lengths are all less than ρ. Describe about each of the centers z_ν the largest circle K_ν lying still entirely within $\mathfrak{G}$. The

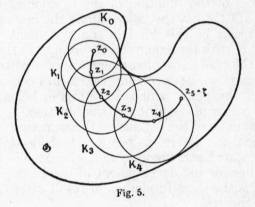

Fig. 5.

radii of these circles are all greater than or equal to ρ. Therefore each of the circles contains the center of the next. We say briefly that the circles form a *circle chain*. We now expand the functions $f_1(z)$ and $f_2(z)$ in power series about each of the centers z_ν, as we did above for $\nu = 0$. In every case, the expansions converge at least in K_ν. We have seen already that they are identical in K_0. Hence, $f_1(z)$ and $f_2(z)$ also coincide at the point z_1 (lying in K_0) and in a *neighborhood* thereof. Consequently (again by the identity theorem for power series) the two expansions coincide in K_1, so that the functions must be equal *at* and in a *neighborhood of* z_2. Therefore they have the same expansions in K_2, etc. The mth step in this argument reads: the functions coincide at $z_m = \zeta$ (and in a neighborhood of ζ). This completes the proof of the theorem.

The method used in this proof is called the **circle-chain method.** This name is suggested by the figure.

In the next chapter we shall concern ourselves in greater detail with the most important consequences of this theorem. Now we consider only a few very simple corollaries.

In order to formulate them conveniently we make use of the following definition.

Definition. *A point z_0 of a region of regularity of the function $f(z)$ is called a* **zero** *of the function if $f(z_0) = 0$. In general, if $f(z_0) = a$, z_0 is called an a-point of $f(z)$.*

We then have

Theorem 1. *Let $f(z)$ be a regular function in $\mathfrak{G}$ and let a be any number. Then $f(z)$ has at most a finite number of a-points in every closed subregion $\mathfrak{G}'$ of $\mathfrak{G}$, unless $f(z)$ is everywhere equal to a.*[1]

Proof: Suppose $f(z)$ had an infinite number of a-points in $\mathfrak{G}'$. These would then have a limit point z_0 situated in $\mathfrak{G}'$ and therefore also in $\mathfrak{G}$. The function which is equal to a at *every* point of the plane is certainly regular everywhere, and in particular in $\mathfrak{G}$. According to the identity theorem, $f(z)$ would have to coincide with this function.

One can state this result in the following form which is often more convenient to apply.

Theorem 2. *If $f(z)$ is regular at z_0, one can describe such a small circle about z_0 as center, that in this circle $f(z)$ never again assumes the value it has at the center unless $f(z)$ has everywhere this same value.*

Theorem 3. *If $f_1(z)$ and $f_2(z)$ are regular in $\mathfrak{G}$, and if both functions, together with all their respective derivatives, coincide for only a single point z_0 of $\mathfrak{G}$, then the functions are identical.*

Proof: If both functions are expanded in power series about the center z_0, identical series are obtained.

---

[1] Or, a limit point of a-points never lies in a region of regularity, but, on the contrary, is necessarily a singular point of $f(z)$. unless $f(z)$ is everywhere equal to a. Or, an infinite number of a-points cannot lie in every neighborhood of a regular point. unless $f(z)$ is everywhere equal to a.

In fact, the coefficients, except for equal numerical factors, are the respective derivatives of the functions at z_0, and hence are equal by hypothesis. Therefore, by the identity theorem, the functions are equal everywhere in $\mathfrak{G}$.

Theorem 4. *If the regular point z_0 is an a-point of the non-constant function $f(z)$, then there is always a definite positive integer α such that the function*

$$f_1(z) = \frac{f(z) - a}{(z - z_0)^\alpha}$$

can, for all points distinct from z_0 of some neighborhood of z_0, be expanded in a power series

$$f_1(z) = b_0 + b_1(z - z_0) + \cdots$$

whose first coefficient is not zero.

Proof: In the expansion $\sum\limits_{n=0}^{\infty} a_n(z - z_0)^n$ of $f(z)$ about the center z_0, $a_0 = a$, and at least one of the succeeding coefficients is not zero. If a_α is the first of these, we have

$$f(z) - a = a_\alpha(z - z_0)^\alpha + a_{\alpha+1}(z - z_0)^{\alpha+1} + \cdots,$$
$$(a_\alpha \neq 0),$$

from which the assertion can be read off. Naturally $b_0 = a_\alpha$ and in general $b_\nu = a_{\alpha+\nu}$, $(\nu = 0, 1, 2, \ldots)$. α is called the **order** of the a-point z_0. Thus every point has a definite (positive integral) order.[1]

Exercises. 1. If the simple closed path C and its interior lie within a region of regularity of $f(z)$, then C encloses only a *finite* number of zeros (more generally: a-points) of $f(z)$.

2. The function $\sin \dfrac{1}{1 - z}$ is regular in the interior of the unit

---

[1] If $f(z)$ is regular at z_0 and $f(z_0) \neq a$, it is often convenient to call the point z_0 an a-point of order zero. According to this, a zero of order zero is a regular point at which the function is not zero.

circle and has there the *infinite number* of zeros $1 - \dfrac{1}{k\pi}$, ($k = 1$, 2, $\cdots$), arising from $\dfrac{1}{1-z} = k\pi$.   Does this contradict Theorem 1 or Exercise 1?   Explain.

3. In connection with §20, Theorem 5, show that at z_0, $|f(z)|$ can have no minimum different from zero, and $\Re(f(z))$ as well as $\Im(f(z))$ can have neither a maximum nor a minimum there.

CHAPTER 8

ANALYTIC CONTINUATION AND COMPLETE DEFINITION OF ANALYTIC FUNCTIONS

§22. The Principle of Analytic Continuation

The considerations of the last chapter culminated in the identity theorem for analytic functions: if two such functions coincide for a neighborhood of a point (or along a small path segment, or only for certain infinite point sets), then they are fully identical. As we have pointed out on p. 86, this implies the strongest constraint for the function: a function is completely determined (i.e., its entire domain of values with all its regular and singular properties) by its values for these point sets.

We shall now be concerned with working out still more clearly the property of analytic functions involved here. To this end we suppose that two functions $f_1(z)$ and $f_2(z)$ are given, of which the first is regular in a region $\mathfrak{G}_1$ and the second is regular in a region $\mathfrak{G}_2$. We further assume that $\mathfrak{G}_1$ and $\mathfrak{G}_2$ have a certain region $\mathfrak{g}$ (however small), but only this region, in common (cf. Fig. 6, where $\mathfrak{g}$ is hatched); and finally, that $f_1(z) = f_2(z)$ everywhere in $\mathfrak{g}$.

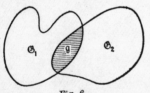

Fig. 6.

Under these conditions the functions f_1 and f_2 determine each other uniquely. In fact, according to the identity theorem, no function other than $f_1(z)$ can be regular in $\mathfrak{G}_1$ and have the same values in $\mathfrak{g}$. Thus, $f_1(z)$ is completely determined by these values in $\mathfrak{g}$ (or what is the same: by $f_2(z)$); and likewise $f_2(z)$ is fully determined by $f_1(z)$.

We can say, therefore, that if two regions $\mathfrak{G}_1$ and $\mathfrak{G}_2$ are in the position just described, and if a regular function is defined in $\mathfrak{G}_1$, then either there is no function at all or *precisely one* function which is regular in $\mathfrak{G}_2$ and coincides with $f_1(z)$ in $\mathfrak{g}$. If such a function $f_2(z)$ exists, then the function $f_1(z)$ defined in $\mathfrak{G}_1$ is said to be *continuable* beyond $\mathfrak{G}_1$ into the region $\mathfrak{G}_2$. When the function $f_2(z)$ has been obtained, $f_1(z)$ is said to have been *continued analytically* into the region $\mathfrak{G}_2$. On the other hand, $f_1(z)$ is the analytic continuation of $f_2(z)$ into the region $\mathfrak{G}_1$. In fact, one has no right to regard $f_1(z)$ and $f_2(z)$ as distinct functions any more. Because of the complete determination of the one by the other, one must regard both as *partial representations* or "elements" of one and the same function $F(z)$ which is regular in the composite region formed by $\mathfrak{G}_1$ and $\mathfrak{G}_2$.

An example will make this clearer. Let $\mathfrak{G}_1$ be the unit circle $|z| < 1$; $\mathfrak{G}_2$ the circle with radius $\sqrt{2}$ and center i, i.e., the circle $|z - i| < \sqrt{2}$. Both circles evidently have a region $\mathfrak{g}$ in common (the reader should make a sketch for himself). In $\mathfrak{G}_1$ let $f_1(z) = \sum_{n=0}^{\infty} z^n$ be given. Is there a function which is regular in $\mathfrak{G}_2$ and coincides with $f_1(z)$ in $\mathfrak{g}$? *If such a function does exist, then there can be only one.* Here $f_2(z) = \dfrac{1}{1-i} \sum_{n=0}^{\infty} \left(\dfrac{z-i}{1-i}\right)^n$ is the required function because this series converges for $\left|\dfrac{z-i}{1-i}\right| < 1$, i.e., for $|z - i| < \sqrt{2}$. and the values of both power series are seen immediately to be equal in $\mathfrak{g}$. This follows from the fact that the sums of both geometric series in their respective circles of convergence can be obtained in closed form and hence compared. (One obtains $\dfrac{1}{1-z}$ in $\mathfrak{g}$ both times.)

$f_1(z)$ and $f_2(z)$ are thus analytic continuations of each other, both are *elements* of one and the same function $F(z)$ which is regular in (at least) the composite region $\mathfrak{G}$ formed by $\mathfrak{G}_1$ and $\mathfrak{G}_2$.

In this simple example we are actually in a position

to obtain the function $F(z)$ in closed form, namely, $F(z) = \dfrac{1}{1-z}$. This is quite impossible in general, however. In fact, $F(z)$ generally can only be calculated by means of its partial representations or elements. Nevertheless, according to §5, $F(z)$ is to be considered a *single* function, the various partial representations *together* furnishing the rule of definition by virtue of which the function $F(z)$ is defined.

We sum up the result, which is called the **principle of analytic continuation,** in the following theorem.

Theorem 1. *Let a regular function $f_1(z)$ be defined in a region $\mathfrak{G}_1$ and let $\mathfrak{G}_2$ be another region which has a certain subregion $\mathfrak{g}$, but only this one, in common with $\mathfrak{G}_1$. Then, if a function $f_2(z)$ exists which is regular in $\mathfrak{G}_2$ and coincides with $f_1(z)$ in $\mathfrak{g}$, there can only be one such function. $f_1(z)$ and $f_2(z)$ are called analytic continuations of each other. They serve as partial representations or elements of one and the same function $F(z)$ determined by them, and $F(z)$ is regular in the composite region formed by $\mathfrak{G}_1$ and $\mathfrak{G}_2$.*

The following questions now arise:

1) If a regular function $f_1(z)$ is defined in a first region $\mathfrak{G}_1$ (e.g., a power series in its circle of convergence), how does one determine whether $f_1(z)$ can be continued into a region $\mathfrak{G}_2$ in the sense just explained, and how is the continuation $f_2(z)$ found?

2) Do other regions $\mathfrak{G}_3$, $\mathfrak{G}_4$, . . . exist, each having a single subregion in common with one of the preceding regions, and are regular functions $f_3(z)$, $f_4(z)$, . . ., respectively, defined therein which constitute continuations, in the sense defined, of the preceding functions?

If so, then all of these functions are uniquely determined by $f_1(z)$ and are therefore to be regarded as elements of *one and the same* function.

3) If one element of a function is given, how does one find all possible further elements, all continuations into adjacent regions?

This comprehensive and apparently very difficult problem admits of a very simple solution, at least theoretically.

Before we present it in §24, let us consider analytic continuation from a somewhat different point of view. In the preceding we have made use of the fact, arising from the expansion theorem, that an analytic function is already determined by its values in a small subregion. Indeed, it is sufficient to know the values only along a small path segment. Accordingly, suppose a path segment k is given in the plane and to every point z of k corresponds a value $\varphi(z)$ of a function. If we consider any region $\mathfrak{G}$ containing k, we are faced with the following alternative: either there is *no function $f(z)$ at all* which coincides with $\varphi(z)$ along k and is regular in $\mathfrak{G}$; or there is *precisely one* such function, and this function is *uniquely determined* by the values along k. In this case we also say that the function defined along k has been continued analytically into the region $\mathfrak{G}$.

In particular, if k is a segment of the real axis, say the interval $x_0 \leq x \leq X$, and if the functional values (which need not be real) corresponding to the points of that segment are denoted by $\varphi(x)$, then we are dealing with the analytic continuation of a (real or complex) function of the *real* variable x. If we have succeeded in continuing the function, $\varphi(x)$ is said to have been continued "into the complex domain." In this connection we can state the following theorem.

Theorem 2. *If it is at all possible to continue a function of the real variable x into the complex domain, then this can be accomplished in only one way.*

The following remarks will place the strong inner constraint of an analytic function in a still clearer light.

Let k be the real segment $0 \leq x \leq \frac{1}{2}$, let the unit circle be the region $\mathfrak{G}$ containing k, and let $\varphi(x)$ be defined on k. If one now considers $\varphi(x)$ on only half the segment, $0 \leq x \leq \frac{1}{4}$, then by the above theorem these functional values already determine whether $\varphi(x)$ can or cannot be continued into the unit circle. In the first case, the values $\varphi(x)$ on the other half of the segment, i.e., on $\frac{1}{4} < x \leq \frac{1}{2}$, are already determined by those on the first half.

Thus, one has *no freedom whatsoever* in the choice of the values $\varphi(x)$ of the function on the second half if one would not make the continuability altogether impossible. One can now apply the same consideration to the first half $0 \leqq x \leqq \frac{1}{4}$, etc. In short, the freedom in the choice of the values of $\varphi(x)$, although not actually illusory, is certainly restricted to a finite number of points, since, according to the identity theorem, the possibility of continuation is already decided by the values of the function at an infinite number of points.

Exercise. Let the real function $F(x)$ be defined by $F(x) = + \sqrt{x^2}$ (i.e., the positive value of $\sqrt{x^2}$) for all real x.

Can this function be continued into the complex domain?

§23. The Elementary Functions

With regard to the last theorem, one can now investigate the more familiar functions of a real variable x to see whether they can or cannot be continued into the complex domain, and discover, in the former case, how the analytic function which furnishes the continuation is constituted.

1. *The rational functions.* Given

$$\varphi(x) = \frac{a_0 + a_1 x + \cdots + a_m x^m}{b_0 + b_1 x + \cdots + b_k x^k},$$

(the a_ν and b_ν are complex), i.e., a rational function, one sees immediately that $\varphi(x)$ is continuable and that

$$f(z) = \frac{a_0 + a_1 z + \cdots + a_m z^m}{b_0 + b_1 z + \cdots + b_k z^k}$$

is the function which continues $\varphi(x)$ into the complex domain. $f(z)$ is regular in the entire z-plane with the exception of those points at which the denominator is zero. (It will be proved in §28, Theorem 3 that there are at most k such points.)

2. e^z, $\sin z$, $\cos z$. The exponential function e^x and the trigonometric functions $\sin x$ and $\cos x$ can be defined by the series

$$e^x = 1 + x + \frac{x^2}{2!} + \cdots + \frac{x^n}{n!} + \cdots = \sum_{n=0}^{\infty} \frac{x^n}{n!},$$

$$\sin x = x - \frac{x^3}{3!} + \frac{x^5}{5!} - + \cdots = \sum_{k=0}^{\infty} (-1)^k \frac{x^{2k+1}}{(2k+1)!},$$

$$\cos x = 1 - \frac{x^2}{2!} + \frac{x^4}{4!} - + \cdots = \sum_{k=0}^{\infty} (-1)^k \frac{x^{2k}}{(2k)!}.$$

If one formally replaces x by z, then each of the resulting series

$$f_1(z) = 1 + z + \frac{z^2}{2!} + \cdots + \frac{z^n}{n!} + \cdots = \sum_{n=0}^{\infty} \frac{z^n}{n!},$$

$$f_2(z) = z - \frac{z^3}{3!} + \frac{z^5}{5!} - + \cdots = \sum_{k=0}^{\infty} (-1)^k \frac{z^{2k+1}}{(2k+1)!},$$

$$f_3(z) = 1 - \frac{z^2}{2!} + \frac{z^4}{4!} - + \cdots = \sum_{k=0}^{\infty} (-1)^k \frac{z^{2k}}{(2k)!},$$

being a power series with $z_0 = 0$, $r = \infty$, represents a function which is regular in the entire z-plane. Since these functions coincide with e^x, $\sin x$, and $\cos x$, respectively, for $z = x$, they are **the** *continuations* of these functions into the complex domain. $f_1(z)$ is therefore called the exponential function and is denoted by e^z; likewise the notations $\sin z$ and $\cos z$ are employed for $f_2(z)$ and $f_3(z)$, respectively. In the following considerations, the properties of these analytic functions are presumed to be familiar to the reader (see *Elem.*, ch. 12). It is now evident from the developments in this chapter that there is *an absolute lack of freedom* in the seemingly arbitrary definition of e^z, $\sin z$, and $\cos z$ for a complex argument as given in the *Elemente*.

They can be defined as regular functions of z only in the manner just shown.

3. The continuations of the functions $\log x$, a^x, $\sqrt[m]{x}$, and others will be investigated after we have formulated the concept of analytic function completely. This will be done in the next paragraph.

§24. Continuation by Means of Power Series and Complete Definition of Analytic Functions

We now proceed to answer questions 1) to 3) which were raised in §22, and shall be able to do so with a single method.

Let the function $f_1(z)$ be defined and regular in $\mathfrak{G}_1$. If z_1 is any point of $\mathfrak{G}_1$, the function can be expanded in a power series about this point as center; thus,

$$(1) \qquad f_1(z) = \sum_{n=0}^{\infty} a_n^{(1)}(z - z_1)^n.$$

Two distinct cases can now occur: the radius of convergence of this series is either $+\infty$ or it has a finite, positive value.

If its radius $r_1 = \infty$, i.e., if the series converges for every z (or *converges everywhere*), then each of the questions can be answered immediately. There *is* a function which continues $f_1(z)$ beyond $\mathfrak{G}_1$; it is regular in the entire plane. Consequently, *no other* function which is regular anywhere can be obtained from $f_1(z)$ by continuation except the one defined by that everywhere-convergent power series.

Example. Let

$$g(z) = 1 - \frac{z^2}{2} - \frac{2}{3!}z^3 - \cdots - \frac{n-1}{n!}z^n - \cdots = -\sum_{n=0}^{\infty} \frac{n-1}{n!}z^n,$$

(this series converges everywhere).

$$h(z) = 1 + z + z^2 + \cdots = \sum_{n=0}^{\infty} z^n,$$

(this series converges only for $|z| < 1$), and set

$$f_1(z) = g(z) \cdot h(z)$$

in the unit circle. No functional values are defined by this formula outside the unit circle. Expanding about the center $z_1 = 0$, one finds upon multiplying out the power series[1]:

$$f_1(z) = \sum_{n=0}^{\infty} \frac{z^n}{n!},$$

which is an expansion of the function valid for the whole plane.

If the radius of convergence r_1 of the expansion (1) has a finite, positive value, choose a point z_2 in the interior of the circle of convergence and distinct from the center. One can then determine the expansion valid for the center z_2:

$$(2) \qquad \sum_{n=0}^{\infty} a_n^{(2)} (z - z_2)^n, \text{ where } a_n^{(2)} = \frac{1}{n!} f_1^{(n)}(z_2).$$

Thus the coefficients can be obtained directly from (1) according to §19, Theorem 5.

Obviously we have

$$(3) \qquad r_2 \geqq r_1 - |z_2 - z_1|$$

for the radius of convergence r_2 of this expansion; i.e., r_2 is at least equal to the distance of the point z_2 from the circumference of the first circle.

If the equality sign holds in (3) (see Fig. 7a), then (2) furnishes the value of the function only for such points at which it was already given by (1). Then the expansion (2) does not give us any new information

[1] $1 - \dfrac{1}{2!} - \dfrac{2}{3!} - \cdots - \dfrac{k-1}{k!} = \dfrac{1}{k!}, \quad (k = 0, 1, 2, \ldots).$

directly. It does show, however, that the point of contact, ζ, of the two circles certainly *cannot* be annexed as a regular point to the first circle of convergence. In other words, it is not possible to cover this point ζ and a neighborhood thereof with functional values in such a manner that a function results which is regular in the enlarged region. Such a point ζ is called a *singular point* on the boundary of the circle of convergence; it is impossible to continue the function over this point. We see then that ζ is a *singular point* for the function $f_1(z)$. If, however, the inequality sign

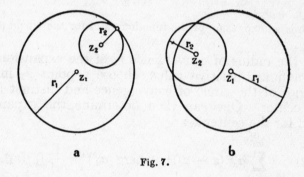

a   b

Fig. 7.

holds in (3) (see Fig. 7b), then the new circle of convergence extends beyond the old one. One has then continued the function over the boundary point ζ of the old circle of convergence in the direction of the radius $z_1 \ldots z_2$. Hence, if a continuation in a radial direction over a boundary point ζ of the first circle of convergence is at all possible, then it is possible to effect it with the aid of these simple power-series expansions.

Now imagine the first functional element to be continued in all possible directions, and likewise suppose the new elements to be continued in all possible directions beyond the newly won domains. Then there arises from the first element a function which is regular in an ever larger domain.

The two following situations are to be noted in this connection.

1. The continuation of the first power series may not be possible in any direction.   Then there is no function which coincides with this power series in its circle of convergence and which is regular in a region which is an enlargement of that circle.   One says that the function is *not continuable*; the circle of convergence is its *natural boundary*.

Example. $f(z) = \sum_{n=1}^{\infty} z^{n!} = z + z^2 + z^6 + \cdots + z^{n!} + \cdots$,

with $r = 1$. If this function $f(z)$ were continuable beyond the unit circle, a certain arc of its circumference would contain only regular points. On every such arc, however, lie an infinite number of points of the form $z_0 = e^{2\pi i \frac{p}{q}}$ with positive integral p and q. If one shows that no point of the form z_0 can even be a point of continuity of $f(z)$, the non-continuability of $f(z)$ will follow. Now, given arbitrarily large (positive integral) g,

$$f(z) = \sum_{n=1}^{q-1} z^{n!} + \sum_{n=q}^{\infty} \rho^{n!}$$

for $z = \rho z_0$ with $0 < \rho < 1$, because $z^{n!} = \rho^{n!}$ for $n \geqq q$. Hence, for $m = 2q + g$,

$$|f(z)| > \sum_{n=q}^{m} \rho^{n!} - \sum_{n=1}^{q-1} |z|^{n!} > (m - q + 1)\, \rho^{m!} - (q - 1).$$

As $\rho \to 1$, the right-hand side approaches $m - 2q + 2 = g + 2$, so that for suitably chosen ρ_0 we must have $|f(z)| > g$ for all $\rho_0 < \rho < 1$. Since g was arbitrary, $|f(z)|$ tends to infinity as z approaches z_0 radially; hence, z_0 cannot be a point of continuity, Q. E. D.

2. The other extreme case, that the power series be continuable beyond the circle of convergence in *all* directions, cannot occur.   For here we have the following important theorem.

Theorem 1.   *At least one singular point of the function defined by a power series exists on the boundary of its circle of convergence.*

Proof: The theorem states that if r_1 is the true radius of convergence of (1), then on the boundary of the circle of convergence there is at least one point ζ over which one cannot continue. We show this by proving that if one can continue over every boundary point ζ of the circle $K: |z - z_1| = r_1$, then r_1 is not the true radius of convergence of (1).

If one can continue over every boundary point ζ of K, then about each of these points as center there is a circle K_ζ, with radius ρ_ζ, into which $f_1(z)$ can be continued. There can be no conflict in the covering of these circles with functional values. If two of these circles have a region in common, then the values of the continuations of $f_1(z)$ into these circles must coincide in that common part, according to the identity theorem, since this common part contains a region lying in K where the coverings are certainly the same. By the Heine-Borel theorem, a finite number of the circles K_ζ are sufficient to cover the entire boundary of K. But these finitely many circles K_ζ, together with K, cover a circular region about the center z_1 with a radius $r > r_1$. Then by the expansion theorem, (1) must converge at least in this larger circle; i.e., r_1 is not the true radius of convergence, Q. E. D.

One is said to continue a given element (in the form of a power series $\Sigma a_n(z - z_0)^n$, say) *along a path* k if the path begins at z_0 and the new center is always chosen on this path.[1] If one supposes such a given element to be continued along all possible paths, then all the points encountered are automatically distributed into two classes: *regular points* and *singular points*, i.e., those which can be included in the interior of a new circle of convergence and those which cannot. To every point z which proves to be regular corresponds a certain functional value w.

We can then make the following definition:

Definition. *The complete analytic function defined by*

[1] More precisely: on that segment of the path which lies between the center and the first point of intersection of the path with the boundary of the circle of convergence

*a given functional element is understood to be the totality
of points which prove to be regular in the course of the
continuation process described above, each covered with its
corresponding functional value.*

The totality of regular points z is called the *region of
existence* or *region of regularity* of this analytic function;
the totality of the corresponding values w is called its
domain of values.

With regard to the gradual growth of the analytic
function from one element, one also speaks of the
analytic configuration, comprising all regular z, each
covered with its corresponding functional value. The
analytic *function* is really the inner bond which unites
each z with its w.

There are still several omissions in this rather complete definition:

a) Agreements will still have to be reached in order
to be able to specify the behavior of a function at
infinity. This will take place in §32.

b) The following situation can occur:

Let us assume that after repeated continuation the
new circle has a region in common with the first one
(in Fig. 8, the fifth of the new circles has the hatched
region in common with the original circle).[1] By virtue
of the new power series, the original functional values
w or else new functional values may correspond to the
points (comprising the hatched region in the figure) of
the old circle of convergence contained in the new one.

In the first case the function is called *single-valued*
(in the region throughout which it has been continued),
otherwise, *multiple-valued*.

c) It is conceivable that an interior (and hence
regular) point of the first circle of convergence prove
to be singular on returning to it in the manner just
described. This *can* actually happen. Thus, the
property of a point of the plane of being regular or

---

[1] The figure rests on the assumption that the original circle of convergence
is the unit circle, that $z = +1$ is the only singular point inside and in a further
neighborhood of that circle, and that the continuation takes place along the
dotted circle $|z - 1| = 1$ in the positive sense.

singular may depend upon the choice of the path or chain of circles used in approaching it.

We must refer the reader to Part II of this *Theory of Functions* for a more accurate examination of the consequences arising from b) and c). In the next para-

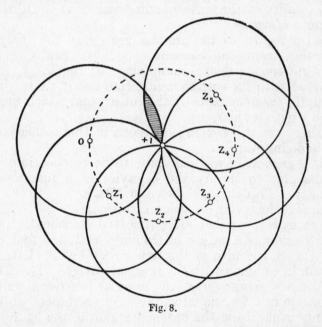

Fig. 8.

graph, however, a theorem will be proved which states that the situation under b) surely *cannot* happen under certain conditions of particularly frequent appearance. The two simplest examples of multiple-valued functions are treated briefly in the paragraph after that.

Exercise. The unit circle is the circle of convergence of the power series $\sum_{n=1}^{\infty} \dfrac{z^n}{n^2}$. Show that the point $+1$ is a singular point of the function represented by the series in the unit circle, by expanding in a new power series with center $z_1 = +\frac{1}{2}$. (Nevertheless, the given series is convergent for $z = +1$!!)

§25. The Monodromy Theorem

Theorem. *Let $\mathfrak{G}$ be a simply connected region and $f_0(z) = \Sigma a_n(z - z_0)^n$ a regular functional element at the point z_0 of $\mathfrak{G}$. Then if $f_0(z)$ can be continued from z_0 along every path within $\mathfrak{G}$, the continuation gives rise to a function which is single-valued and regular in the entire region $\mathfrak{G}$.*

We observe beforehand that every element obtained by continuation, in which only power series are used, converges at least in the largest circle (about the center of the element) which does not project beyond $\mathfrak{G}$. For, on the boundary of its true circle of convergence there is at least one singular point, which obstructs the continuation. By hypothesis, such an obstruction does not occur anywhere in the interior of $\mathfrak{G}$.

We have to show, evidently, that if one continues $f_0(z)$ from z_0 to z_1 along two different paths k_1 and k_2 lying within $\mathfrak{G}$, then one obtains the *same* element $f_1(z) = \Sigma b_n(z - z_1)^n$ at z_1 both times. Since, in short, the continuation process proceeds quite uniquely back and forth,[1] we can also say that if one continues $f_0(z)$ from z_0 to z_1 along k_1 and continues the element $f_1(z)$ obtained at z_1 back to z_0 along k_2, then one obtains once more the initial element $f_0(z)$ at z_0. It suffices then to show that the continuation of an element along a closed path within $\mathfrak{G}$ leads back to this same element. We prove this indirectly by showing that if the continuation of an element along a closed path C lying within $\mathfrak{G}$ does *not* lead back to this element, then this contradicts the hypothesis that our continuations are possible along *every* path within $\mathfrak{G}$. A finite number of centers ζ_0, $\zeta_1, \ldots, \zeta_m$ on the path are required for the continuation along C, beginning at ζ_0, say. Each lies in the circle of convergence about its predecessor and its successor

---

[1] One has only to imagine the successive centers to be chosen so that each lies in the circle of convergence about the preceding center and *the succeeding center.*

if the distance between any two successive ones is chosen to be smaller than the distance of the path C from the boundary of the region. Hence, if one replaces C by the polygon p with the vertices $\zeta_0, \zeta_1, \ldots, \zeta_m,$ the continuations along C and p are exactly the same. Our continuations along p, then, also do not lead back to the initial element. Now, either p is simple, or, by Lemma 1, can be decomposed into a finite number of simple closed polygons and a finite number of segments described twice, once in each direction. In any case, there is at least one simple closed subpolygon of p; for if p only contained segments described twice, our continuations along p would necessarily have to return to the initial element. There must be, then, a simple closed subpolygon p' of p along which the continuations, proceeding in the positive sense, do not lead back to the initial element.

Let us decompose p' into two subpolygons by means of a diagonal lying within p' (and hence within $\mathfrak{G}$). The continuations along one of the subpolygons (in the positive sense) do not lead back to the initial element, since one continues back and forth along the diagonal. By further subdividing this polygon, one must eventually arrive at a triangle along which the continuations do not return to the initial element. If one decomposes this triangle as in the proof of Cauchy's integral theorem (see Fig. 1), one obtains a sequence of nested triangles, which close down on a point ζ, along each of which the continuations do not lead back to the initial element. This is impossible, however. For, the element with center ζ has a positive radius ρ. As soon as the diameter of one of the triangles containing the point ζ is less than ρ, the continuation around this triangle must surely return to the initial element, since in this process one does not have to go beyond the circle with radius ρ and center ζ, every point of which is covered with one regular functional value. This proves the monodromy theorem.

§26. Examples of Multiple-valued Functions

The effective calculation of the entire analytic configuration, that is, the separation of all z into regular and singular points and the association of the functional values with the regular z, cannot, in general, be accomplished by the given method.   Its value consists chiefly in giving an insight into the nature of the matter; it has merely the character of an existence theorem.

The following two examples show how entirely different means lead to the objective in particular cases.

1. $w = f(z) = \log z.$

We have already discovered in §14, 6 that

$$f(z) = \int_1^z \frac{d\zeta}{\zeta}$$

is a regular analytic function in the right half-plane, provided the path of integration is also confined to this half-plane.   Since the natural logarithm can be defined for $x > 0$ by

$$\log x = \int_1^x \frac{d\xi}{\xi},$$

it is immediately evident that $f(z)$ is the analytic continuation of $\log x$ into the complex domain, because $f(z) = \log x$ for $z = x > 0$.

What is the domain of existence of $f(z)$ and what is its domain of values?

The integral for $f(z)$ always has a meaning if the path of integration avoids the origin.   Hence (see §14, Theorem 3), the function $f(z)$ is regular everywhere except at the origin.[1]

---

[1] This is true in the finite part of the plane.   After reading §32, however, which treats of the behavior of an analytic function at infinity, the reader will be able to verify that the point ∞ is a branch-point (defined below) of infinite order of the function $\log z$, and a branch-point of order $m - 1$ of the function appearing in the next example.

It is not single-valued, however. In order to find, for example, $f(-1) = \log(-1)$, one can first choose the upper half and then the lower half of the unit circle as the path of integration. One obtains (cf. §10, Example 1)

$$+ \pi i, \ - \pi i, \text{ respectively,}$$

which is in agreement with the fact that the integral taken over the whole unit circle in the positive sense is equal to $2\pi i$.

According to Cauchy's theorem, the integral has the same values if any other path lying entirely within the upper half-plane (lower half-plane) is chosen.

If, however, one chooses a path which begins at $+1$ and encircles the origin m times in the positive sense before terminating in -1, one obtains (see §10, 1)

$$\log(-1) = \pi i + 2m\pi i,$$

since the integral taken along a path which encloses the origin once is equal to $2\pi i$. Likewise, by encircling the origin m times in the negative sense, one obtains

$$\log(-1) = -\pi i - 2m\pi i.$$

Thus, depending on the choice of the path, we obtain an infinite number of values for $\log(-1)$, all having the form

$$\log(-1) = \pi i + 2k\pi i; \quad (k = 0, \pm 1, \pm 2, \dots).$$

It is easy to see that, according to Cauchy's theorem, one obtains one of these values using *any* path extending from $+1$ to -1. What holds for the point -1 naturally holds for every other point.

We can say, then, that the function $\log z$ is regular in the entire finite plane with the exception of the origin. It is infinitely multiple-valued, but in such a manner, that all values of $\log z$ for a particular z can be obtained from one of them by the addition of an arbitrary integral multiple of $2\pi i$. Each of these infinitely many values of $\log z$ is called a *determination of the logarithm* at the point z. Each of these determinations constitutes a single-valued, regular function in a neighborhood of every point different from zero, or, more generally, in every simply connected region ⑤ which does not contain the origin. The single-valued functional element which is thereby selected from the whole domain of values of log z is also called a *branch* of the multiple-valued function. In §20 we developed such a branch (actually the so-called *principal value*) of log z in a power series for a neighborhood of + 1.

One can also develop the same properties of log z, though not as conveniently, by applying the general methods of the preceding paragraph to this power series as the initially given functional element. In particular, one can show directly that if one continues the power series just mentioned once around the origin in the positive sense in a manner similar to that sketched in Fig. 8 (always choosing the new centers on the unit circle, let us say), one does *not* return with the principal value to the initial circle. On the contrary, the functional values have increased by $2\pi i$. The origin, in the neighborhood of which log z is not single-valued (and which is the only finite singular point of log z), is consequently called a branch-point or winding-point of log z. In this case the branch-point is of *infinite order*.

We presume the elementary properties of the function log z to be familiar to the reader (see *Elem.*, ch. 13), and only emphasize once more that the ambiguity of log z, which appears to be rather arbitrary in some presentations, is actually an *essential* property of this function. It arises with *absolute necessity* from each of its elements, no matter how they be given, on the basis of the continuation principle.

For each of the infinitely many determinations of log z we have $e^{\log z} = z$.

2. $$w = f(z) = \sqrt[m]{z}.$$

The real function $\sqrt[m]{x}$, defined and positive for

$x > 0$, can also be continued into the complex domain. For,

$$f(z) = e^{\frac{1}{m} \log z}$$

is, with $\log z$, a function which is regular in the entire (finite) z-plane with the exception of the origin, though not single-valued in a neighborhood of the origin. However, if we choose a simply connected region $\mathfrak{G}$ which does not contain the origin, e.g., the entire plane exclusive of the real numbers less than or equal to zero,[1] then every branch of $\log z$ is a single-valued, regular function there.

In particular, let us select that branch which has the value zero for $z = +1$, and hence is equal to the real value $\log x$ for all $x > 0$, and denote this so-called principal value by $\operatorname{Log} z$. Then the function

$$f_0(z) = e^{\frac{1}{m} \operatorname{Log} z},$$

which is regular in $\mathfrak{G}$, is the required continuation of the positive real function $\sqrt[m]{x}$; for, $f_0(x) = e^{\frac{1}{m} \log x} = x^{\frac{1}{m}} = \sqrt[m]{x}$. We therefore denote the function $f(z)$ by $\sqrt[m]{z}$; $f_0(z)$ is called the principal value of $\sqrt[m]{z}$.

According to this definition, the function $\sqrt[m]{z}$ at first appears to be infinitely multiple-valued; it is, however, only m-valued. For, all values of $\log z$ are contained in

$$\log z = \operatorname{Log} z + 2k\pi i, \quad (k = 0, \pm 1, \pm 2, \ldots),$$

so that

$$f(z) = \sqrt[m]{z} = e^{\frac{1}{m} \operatorname{Log} z} \cdot e^{\frac{2k\pi i}{m}} = e^{\frac{2k\pi i}{m}} \cdot f_0(z).$$

The factor before $f_0(z)$ can only take on m distinct

---

[1] This region is said to be the plane "*cut*" *along the negative real axis.*

values,[1] because two values of k which differ only by a multiple of m give it the same value. The m branches of $\sqrt[m]{z}$ consequently differ from the principal branch only by constant factors. We allow k to assume the values $0, 1, 2, \ldots, m - 1$ and accordingly obtain as representations of the m branches:

$$f_k(z) = e^{\frac{2k\pi i}{m}} \, e^{\frac{1}{m} \operatorname{Log} z}, \quad k = 0, 1, 2, \ldots, m - 1.$$

We have derived these results:

1) $\sqrt[m]{x}$ can be continued into the complex domain.

2) The analytic function $\sqrt[m]{z}$, which is thereby uniquely determined, is regular in the entire finite plane except at the origin.

3) It is m-valued. The origin is the only finite branch-point, and it is of order $m - 1$.[2] By continuing analytically around this point, the function is multiplied by an mth root of unity. We have always $\left(\sqrt[m]{z} \right)^m = z$.

We presume, again, that the elementary properties of the function $\sqrt[m]{z}$ are familiar to the reader, so that we may be content with this brief exposition of its analytic structure.

Exercises. 1. Expand the principal value of $\sqrt[m]{z}$ in a power series for a neighborhood of the point $+ 1$; in particular, for $m = 2$.

2. The function a^z, where a is an arbitrary complex constant (different from zero and unity) is defined by the relation

$$a^z = e^{z \log a}.$$

Where is this function regular? Is it single-valued or multiple-valued? Accordingly, can a^z be single-valued? What is the meaning of i^i ?

---

[1] These are the m distinct mth roots of unity, since $\left(e^{\frac{2k\pi i}{m}} \right)^m = e^{2k\pi i} = + 1$.

[2] It is said to be of order $m - 1$ because obviously the first stage of ambiguity occurs for $m = 2$.

CHAPTER 9

ENTIRE TRANSCENDENTAL FUNCTIONS

§27. Definitions

According to the developments of the preceding chapter, the simplest functions appear to be those whose power-series expansions converge in the entire plane; for, such a function is regular in the whole plane, and its power-series expansion, which we may now assume to be in the form

$$w = f(z) = \sum_{n=0}^{\infty} a_n z^n,$$

furnishes for every z the corresponding value of the function. These functions therefore are necessarily single-valued. They are called, briefly, **entire functions**[1] and are classified as *entire transcendental functions* and *entire rational functions* (or *polynomials*) according as an infinite number or only a finite number, respectively, of the coefficients a_n of the expansion are different from zero. In the latter case, if a_m is the last non-zero coefficient, m is called the *degree* of the polynomial. e^z, sin z, and cos z, for example, are entire transcendental functions.

The theorems of the following paragraph deal with the characteristic behavior of these functions. If $f(z)$ has one and the same value c for all z, then, to be sure, $f(z)$ is also an entire function: a polynomial of degree zero. It represents a degenerate form, however, to which the following theorems do not apply.

§28. Behavior for Large $|z|$

1. We begin with the so-called **first Liouville theorem.**

[1] Or, by some authors, "integral functions "

Theorem 1. *A non-constant entire function assumes arbitrarily large values outside every circle; i.e., if R and G are arbitrary (large) positive numbers, then points z exist for which*

$$|z| > R \quad and \quad |f(z)| > G.$$

Proof: We prove the theorem in the equivalent form: *A bounded[1] entire function necessarily reduces to a constant.* In fact, if a constant M exists such that $|f(z)| \leq M$ for all z, then it follows immediately from Cauchy's inequality $|a_n| \leq \dfrac{M}{\rho^n}$ that $a_n = 0$, for $n = 1$, 2, . . ., because any arbitrarily large number may be substituted for ρ. Hence $f(z) \equiv a_0$.

2. If, in particular, the function in question is an entire *rational* function, i.e., a *polynomial*, Theorem 1 can be sharpened to the following result.

Theorem 2. *If $f(z)$ is a polynomial of degree m, ($m \geq 1$), and G is an arbitrary positive number, then R can be assigned so that $|f(z)| > G$ for all $|z| > R$.*

Proof: We have

$$f(z) = a_0 + a_1 z + a_2 z^2 + \cdots + a_m z^m$$

$$= z^m \left[a_m + \frac{a_{m-1}}{z} + \cdots + \frac{a_0}{z^m} \right].$$

Hence, if we set $|z| = r$,

$$|f(z)| \geq r^m \left[|a_m| - \frac{|a_{m-1}|}{r} - \cdots - \frac{|a_0|}{r^m} \right],$$

which, since $a_m \neq 0$, is larger than $\frac{1}{2} |a_m| r^m$, hence, larger than G, and in fact, greater than $G r^{m-1}$, for all sufficiently large r.

3. A very simple proof of the **fundamental theorem of algebra** (cf. *Elem.*, §39) results from these theorems.

---

[1] A *function* is said to be bounded in a region if the domain of values of the function for that region is a bounded set of numbers.

Theorem 3. *If $f(z)$ is a polynomial of degree m, $(m \geqq 1)$, then the equation $f(z) = 0$ has at least one solution.* Briefly: $f(z)$ has zeros.

Proof: If we had $f(z) \neq 0$ for all z, then $\dfrac{1}{f(z)} = g(z)$ would also be an entire (non-constant) function. Hence, by Liouville's theorem there would be points z outside of every circle, for which

$$| g(z) | > 1, \quad \text{that is,} \quad | f(z) | < 1,$$

contradicting Theorem 2 just proved.

An entire *transcendental* function need not have any zeros; e^z, for example, is an entire function with no zeros.

4. If, on the other hand, we are concerned with an entire *transcendental* function in connection with Liouville's theorem, then the latter can be sharpened to the following result.

Theorem 4. *If $f(z)$ is an entire transcendental function, and if the numbers $G > 0$, $R > 0$, and $m > 0$ are given arbitrarily, there always exist points z for which*

$$| z | > R \quad and \quad | f(z) | > G \cdot | z |^m.$$

Proof: We prove this theorem, as we did Theorem 1, in an equivalent form: If $f(z)$ is an entire function, and if two positive constants M and m exist such that

$$| f(z) | \leqq M | z |^m$$

for all z, then $f(z)$ is a *polynomial* of degree less than or equal to m. In fact, the inequality $| a_n | \leqq M \rho^{-n+m}$ now holds for all ρ. Hence, we must have $a_n = 0$ for $n > m$.

5. The remarkable **Casorati-Weierstrass theorem** follows from all these theorems.

Theorem 5. *Outside every circle, an entire transcendental function comes arbitrarily close to every value.*

Or in symbols: if the complex number c and the positive numbers ϵ and R are given arbitrarily, then the inequality

$$|f(z) - c| < \epsilon$$

is satisfied by suitable $|z| > R$.[1]

Proof: a) If $f(z)$ has an *infinite number* of c-points, then according to §21, Theorem 1 they cannot all lie in the circle $|z| \leq R$; so that in the exterior of this circle the equation $f(z) - c = 0$ actually has solutions.

b) If $f(z)$ has *no* c-points, then $\dfrac{1}{f(z) - c} = f_1(z)$ also is a non-constant entire function, so that according to Theorem 1, points z, with $|z| > R$, can be determined such that $|f_1(z)| > \dfrac{1}{\epsilon}$; i.e., $|f(z) - c| < \epsilon$.

c) If $f(z)$ has a *finite number* of c-points, let these be $z_1, z_2, \ldots, z_k$ of orders $\alpha_1, \alpha_2, \ldots, \alpha_k$, respectively. Then (see §21, Theorem 4)

$$\frac{f(z) - c}{(z - z_1)^{\alpha_1}(z - z_2)^{\alpha_2} \cdots (z - z_k)^{\alpha_k}} = f_1(z)$$

is also an entire function, but one *with no* zeros, so that $\dfrac{1}{f_1(z)} = f_2(z)$ is an entire and, indeed, a *transcendental* function. Hence, by Theorem 4, the inequality

$$|f_2(z)| > \frac{2}{\epsilon} \cdot |z|^m$$

is satisfied outside *every* circle for certain z. Let m here be equal to $\alpha_1 + \alpha_2 + \cdots + \alpha_k$. Then

$$(1) \qquad |f(z) - c| < \epsilon \left| \frac{(z - z_1)^{\alpha_1} \cdots (z - z_k)^{\alpha_k}}{z^m} \right|$$

---

[1] In other words: no matter how large R is prescribed, the set of values w assumed by $f(z)$ in the exterior of the circle $|z| = R$ is everywhere dense in the w-plane.

Since

$$(2) \qquad \left| \frac{(z - z_1)^{\alpha_1} \cdots (z - z_k)^{\alpha_k}}{z^m} \right| < 2$$

for all sufficiently large z, say for all $|z| > R_1 > R$, it follows, if we also suppose that $|z| > R_1$ in (1), that the relations (1) and (2) hold for these certain z, so that

$$|f(z) - c| < \epsilon$$

is also satisfied.

Exercise. Prove the last theorem more simply and quickly with the aid of the Laurent expansion of

$$\frac{1}{f(z) - c}$$

for large $|z|$, treated in §§29 and 30.

SINGULARITIES

CHAPTER 10

THE LAURENT EXPANSION

§29. The Expansion

Up to now we have examined functions exclusively in domains in which they are regular. We shall now consider the case that there are singular points in the interior of the domain; the function is assumed to be single-valued there. In order to have something definite before us, let us assume that $f(z)$ is single-valued and regular in a concentric annular ring with center z_0, whereas nothing is known about the behavior of the function outside the larger circle K_1 with radius r_1 and inside the smaller circle K_2 with radius r_2

$$(0 < r_2 < r_1).$$

We shall then obtain an expansion which converges and represents $f(z)$ for every z in the ring, i.e., for every z such that $r_2 < |z - z_0| = \rho < r_1$. To this end, choose two radii ρ_1 and ρ_2 for which

$$r_2 < \rho_2 < \rho < \rho_1 < r_1.$$

Let the circles having these radii and the center z_0 be C_1 and C_2, respectively. $f(z)$ then is regular within and on the boundary of the ring between these circles, since this ring lies entirely within the first ring. Connect C_1 and C_2 by means of two radial auxiliary paths k' and k'' which do not pass through z. Proceeding exactly as in §14, 4 we obtain

$$f(z) = \frac{1}{2\pi i} \int_{C_1} \frac{f(\zeta)}{\zeta - z} d\zeta - \frac{1}{2\pi i} \int_{C_2} \frac{f(\zeta)}{\zeta - z} d\zeta,$$

if C_1 and C_2 are both oriented positively. Now (in this connection see the proof of Theorem 1 in §20)

a) for the first integral, since ζ here is a point of the circle C_1,

$$\frac{1}{\zeta - z} = \frac{1}{\zeta - z_0} \frac{1}{1 - \dfrac{z - z_0}{\zeta - z_0}} = \sum_{n=0}^{\infty} \frac{(z - z_0)^n}{(\zeta - z_0)^{n+1}},$$

a series which converges uniformly for all ζ on C_1 because $\left| \dfrac{z - z_0}{\zeta - z_0} \right| < \dfrac{\rho}{\rho_1} < 1;$

b) for the second integral, since ζ here lies on C_2,

$$\frac{1}{\zeta - z} = -\frac{1}{z - z_0} \cdot \frac{1}{1 - \dfrac{\zeta - z_0}{z - z_0}} = -\sum_{n=0}^{\infty} \frac{(\zeta - z_0)^n}{(z - z_0)^{n+1}},$$

a series which converges uniformly for all ζ on C_2 because $\left| \dfrac{\zeta - z_0}{z - z_0} \right| = \dfrac{\rho_2}{\rho} < 1.$ If these special expansions of $\dfrac{1}{\zeta - z}$ are substituted in the respective integrals, the integrations may be carried out term by term because of the uniform convergence with respect to ζ, and we obtain

$$f(z) = \sum_{n=0}^{\infty} \frac{1}{2\pi i} \int_{C_1} \frac{f(\zeta)}{(\zeta - z_0)^{n+1}} (z - z_0)^n d\zeta$$

$$+ \sum_{n=0}^{\infty} \frac{1}{2\pi i} \int_{C_2} \frac{f(\zeta)(\zeta - z_0)^n}{(z - z_0)^{n+1}} d\zeta.$$

If, for abbreviation, we set

$$\frac{1}{2\pi i} \int_{C_1} \frac{f(\zeta)}{(\zeta - z_0)^{n+1}} d\zeta = a_n, \qquad (n = 0, 1, 2, \ldots),$$

and

$$\frac{1}{2\pi i}\int_{C_2} f(\zeta)(\zeta - z_0)^{n-1}d\zeta = \frac{1}{2\pi i}\int_{C_2} \frac{f(\zeta)d\zeta}{(\zeta - z_0)^{-n+1}} = a_{-n},$$

$$(n = 1, 2, \ldots),$$

we have

$$f(z) = \sum_{n=0}^{\infty} a_n(z - z_0)^n + \sum_{n=1}^{\infty} a_{-n}(z - z_0)^{-n},$$

which is usually written more briefly as

$$f(z) = \sum_{n=-\infty}^{+\infty} a_n(z - z_0)^n.$$

We have thus obtained a representation of $f(z)$ as the sum of a power series Σ_1 of ascending powers of $z - z_0$ and a power series Σ_2 of descending powers of $z - z_0$. Both series converge if z lies in the interior of the annular region between K_1 and K_2. For, it is clear that the values of a_n and a_{-n} are independent of the form of the paths of integration of the integrals defining those coefficients, and hence, of ρ_1, ρ_2, respectively. According to §14, 4, any other closed path lying entirely within the annular region between K_1 and K_2 and encircling K_2 once may be chosen instead of C_1, C_2, respectively. The series obtained is called the **Laurent expansion of** $f(z)$ for the annular region.

§30. Remarks and Examples

In order to understand thoroughly the formula of the preceding paragraph, we consider separately the functions represented by the two sums Σ_1 and Σ_2.

$$f_1(z) = \Sigma_1 = \sum_{0}^{\infty} a_n(z - z_0)^n$$

is an ordinary power series in $z - z_0$. Consequently, it converges for *all* z within K_1, and represents a regular function there.

$$f_2(z) = \Sigma_2 = \sum_{n=1}^{\infty} a_{-n}(z - z_0)^{-n}$$

likewise proves to be an ordinary power series; one has only to set

$$a_{-n} = b_n \quad \text{and} \quad (z - z_0)^{-1} = z',$$

whereupon

$$f_2(z) = \sum_{n=1}^{\infty} b_n z'^n.$$

Since Σ_2 certainly converges for $r_2 < |z - z_0| < r_1$, this new series certainly converges for

$$\frac{1}{r_1} < |z'| < \frac{1}{r_2}.$$

Hence, since it is an ordinary power series in z', it converges for *all* $|z'| < \frac{1}{r_2}$, and represents a regular function of z' there. Returning to z, this means that Σ_2 converges for *all* z for which

$$|z - z_0| > r_2,$$

i.e., everywhere outside of K_2, and represents a regular function of z there. $f(z)$ is thus decomposed into two functions, one regular within K_1 and the other regular without K_2. *Both* are regular in the annular region.

From this and the uniqueness of the Laurent expansion, which will be proved immediately, it follows at once that the exact region of convergence of the same is the broadest ring which can be formed from the hitherto existing ring by concentric contraction of the inner circle K_2 and expansion of the outer circle K_1

and which is still devoid of singular points. There is, therefore, at least one singular point on each of the two circles bounding the ring. (If there is no singular point at all in the interior of K_2, then the inner region, and with it, f_2, Σ_2 would be entirely eliminated by this process.)

The Laurent expansion just found is the only one possible, just like the Taylor expansion. For, assume that

$$f(z) = \sum_{n=-\infty}^{+\infty} a_n(z - z_0)^n \quad \text{and} \quad f(z) = \sum_{n=-\infty}^{+\infty} c_n(z - z_0)^n$$

are simultaneously *valid for a common annular region*. Multiply both expansions by $(z - z_0)^{-k-1}$ and integrate along a circle with center z_0 lying entirely within the annular region, so that the resulting series converges uniformly on that circle with respect to z. It follows that

$$2\pi i a_k = 2\pi i c_k, \quad \text{that is,} \quad a_k = c_k,$$
$$(k = 0, \pm 1, \pm 2, \ldots).$$

Examples. The following expansions are found without difficulty:

(1) $\dfrac{1}{(z - 1)(z - 2)} = -\sum_{n=0}^{\infty} \dfrac{z^n}{2^{n+1}} - \sum_{n=1}^{\infty} \dfrac{1}{z^n}, \quad (1 < |z| < 2),$

or

$$\frac{1}{(z - 1)(z - 2)} = \sum_{n=2}^{\infty} \frac{2^{n-1} - 1}{z^n}, \quad (2 < |z| < \infty).$$

Here we have two different expansions for the *same* function. However, this does not contradict the theorem just proved, since the expansions are valid for *different annular regions*.

(2) $e^z + e^{\frac{1}{z}} = 2 + \sum_{n=1}^{\infty} \dfrac{z^n}{n!} + \sum_{n=1}^{\infty} \dfrac{1}{n!}\dfrac{1}{z^n} = 1 + \sum_{n=-\infty}^{+\infty} \dfrac{z^n}{|n|!},$

$$(0 < |z| < \infty),$$

(3) $\sin\dfrac{1}{z - 1} = \dfrac{1}{z} + \dfrac{1}{z^2} + \dfrac{5}{6}\dfrac{1}{z^3} + \dfrac{1}{2}\dfrac{1}{z^4} + \cdots, (1 < |z| < \infty).$

Exercise.   Expand the functions

$$\frac{1}{e^{1-z}} \quad \text{for} \quad |z| > 1$$

and

$$\sqrt{(z-1)(z-2)} \quad \text{for} \quad |z| > 2$$

in Laurent series.

CHAPTER 11

THE VARIOUS TYPES OF SINGULARITIES

§31. Essential and Non-essential Singularities or Poles

The case that the only singular point of $f(z)$ in the interior of K_2 is the center z_0 deserves special consideration. The Laurent expansion

$$(1) \qquad f(z) = \sum_{n=-\infty}^{+\infty} a_n(z - z_0)^n$$

converges then for all z for which $0 < |z - z_0| < r_1$, where $r_1 \, (> 0)$ is the distance from z_0 to the nearest singular point. In this case, z_0 is called an *isolated singularity*, and an expansion of the form (1)′ always exists in a neighborhood of such an isolated point if $f(z)$ is single-valued there. If that part of the expansion (1) containing the descending powers of $z - z_0$ is again (see above) written in the form $\Sigma b_n z'^n$, it is evident that in this case it represents an *entire* function of z'. According as this entire function is an entire transcendental or an entire rational function, i.e., according as that part of the expansion involving the descending powers of $z - z_0$ contains an infinite number or only a finite number of terms (but then at least one), z_0 is called an *essential* or a *non-essential singularity*. In the latter case, z_0 is also called briefly a **pole**. If $a_{-m} \, (m \geq 1)$ is the last coefficient which is not zero, z_0 is called a pole of order m; multiplication by $(z - z_0)^m$ (but by no smaller power) transforms $f(z)$ into a function which is regular at z_0 and in a neighborhood thereof, and which is different from zero at z_0.

The terms "pole" and "essential singularity" apply only to isolated singular points in whose neighborhood the function is single-valued (see p. 103). That part

123

of the expansion containing the descending powers of $z - z_0$ is called the *principal part* of the function at z_0. The following theorems bear out the great difference in the character of the two kinds of singularities.

Theorem 1. *If $f(z)$ has a pole at z_0 (that is, if $\Sigma_2 = \Sigma b_n z'^n$ is an entire rational function of z') and if $G > 0$ is given arbitrarily, then it is possible to assign a $\delta > 0$ such that*

$$| f(z) | > G$$

for all $| z - z_0 | < \delta$; i.e., $f(z)$ is very large in absolute value for all z lying close to z_0; or, as a pole is approached the function becomes definitely infinite. (In this connection cf. §28, 2.)

Proof: Let z_0 be a pole of order α, so that

$$f(z) = \frac{a_{-\alpha}}{(z - z_0)^\alpha} + \cdots + a_0 + a_1(z - z_0) + \cdots$$

$$= \frac{a_{-\alpha}}{(z - z_0)^\alpha} \left\{ 1 + b_1(z - z_0) + \cdots \right\},$$

(with $a_{-\alpha} \neq 0$, $b_k = \dfrac{a_{-\alpha+k}}{a_{-\alpha}}$, $k = 1, 2, \ldots$).

Choose δ so small that $\delta^\alpha < | a_{-\alpha} | / 2G$ and that the absolute value of the expression in the braces is greater than $\frac{1}{2}$ for all $| z - z_0 | < \delta$. This is certainly possible since we are dealing with a power series with the constant term $+ 1$. Then for all $| z - z_0 | < \delta$ we have

$$|f(z) | \geqq \frac{| a_{-\alpha} |}{\delta^\alpha} \cdot \frac{1}{2} > G, \qquad \text{Q. E. D.}$$

2. The following analogue of Theorem 5 in §28 is also called the **Casorati-Weierstrass theorem.**

Theorem 2. *If $f(z)$ has an essential singularity at z_0 (that is, if $\Sigma_2 = \Sigma b_n z'^n$ is an entire transcendental function* *of z'), then $f(z)$ in every neighborhood of z_0 comes arbitrarily close to every number.   More precisely: if c is an arbitrary complex number and δ and ϵ are two arbitrary (small) positive numbers, then points z always exist for which*

$$|z - z_0| < \delta \quad and \quad |f(z) - c| < \epsilon. \,^1$$

Proof: Admitting the constant term to the second sum we set

$$f(z) = \sum_{n=1}^{\infty} a_n(z-z_0)^n + \sum_{n=0}^{\infty} a_{-n}(z-z_0)^{-n} = \varphi_1(z) + \varphi_2(z).$$

$\varphi_1(z)$ is continuous at z_0 and $\varphi_1(z_0) = 0$. Hence, $\delta_1 \leqq \delta$ can be assigned so that $|\varphi_1(z)| < \frac{1}{2}\epsilon$ for *all* $|z - z_0| < \delta_1$. $\varphi_2(z) = \sum_{n=0}^{\infty} b_n z'^n$, on the other hand, is an entire *transcendental* function of z', so that by the Casorati-Weierstrass theorem in §28 the condition $|\varphi_2(z) - c| < \frac{1}{2}\epsilon$ is satisfied for certain very large z', e.g., such for which $|z'| > 1/\delta_1$. This means that $|\varphi_2(z) - c| < \frac{1}{2}\epsilon$ for certain z with $|z - z_0| < \delta_1$.

For these z, then,

$$|f(z) - c| \leqq |\varphi_1(z)| + |\varphi_2(z) - c| < \epsilon, \qquad \text{Q. E. D.}$$

Examples.

1. $e^{\frac{1}{z}}$ has an essential singularity at $z = 0$ (cf. §30, Example 2).
2. A rational function

$$f(z) = \frac{a_0 + a_1 z + \cdots + a_m z^m}{b_0 + b_1 z + \cdots + b_k z^k}$$

can be singular only at those points at which the denominator is zero.   Let z_1 be a zero of order α of the denominator and at the same time a zero of order β of the numerator ($\alpha \geqq 0$, $\beta \geqq 0$; cf. p. 90, footnote).   Then it is easy to see that $f(z)$ has a pole of order $\alpha - \beta$ at z_1 if $\alpha > \beta$, a zero of order $\beta - \alpha$ if $\beta \geqq \alpha$.[2]   (This

---

[1] In other words: no matter how small $\delta > 0$ is prescribed, the set of values w assumed by $f(z)$ in the interior of the circle $|z - z_0| < \delta$ is everywhere dense in the w-plane.

[2] In this case $f(z_1)$ is to be defined as the value $\lim_{z \to z_1} f(z)$.

example shows already that it will be advantageous to regard poles as zeros of negative order.) Thus, at any assignable distance from the origin a rational function has no other singularities than poles (cf. §32, Examples 2 and 3 and Theorem 1 in this connection).

3. The functions tan z and cot z are discontinuous, and therefore singular, at the zeros of cos z, sin z, respectively. It is easily seen that the singularities there are poles of the first order.

Let us investigate cot z at the point $z = 0$. This point, in any case, is an isolated singularity, since the nearest new zeros of sin z are $z = \pm \pi$. Consequently, cot z admits of a Laurent expansion which one knows in advance must necessarily be valid for all z for which

$$0 < |z| < \pi$$

and only these z. If one proceeds to carry out the division of the power series for cos z and sin z (cf. *Elem.*, §43), the beginning of the expansion is found to be

$$\cot z = \frac{1}{z} - \frac{1}{3} z - \frac{1}{45} z^3 - \cdots.$$

Because of the uniqueness of such an expansion (see §30), this is the Laurent expansion of cot z for the neighborhood of the point $z = 0$. From it we read off immediately that $z = 0$ is a pole of the first order.

We shall not enter into an investigation of non-isolated singularities and singular points in whose neighborhood the function is not single-valued (such as $z = 0$ for log z and for $\sqrt[m]{z}$). Concerning the latter cf. ch. 4 of *Theory of Functions* II.

Exercise. Verify the validity of the Casorati-Weierstrass theorem for the function $e^{1/z}$ by investigating the values which it assumes in the neighborhood of the origin on the radii emanating from that point. Determine the points z at which $e^{1/z} = i$. What sort of point set do these constitute?

§32. Behavior of Analytic Functions at Infinity

There is an omission in our definition of the complete analytic function (§24); we still have to reach agreements as to how to describe the behavior of a function at infinity. As before, we confine ourselves to the case that $f(z)$ is single-valued and regular in a neighborhood of the point ∞ (see §2). Let $f(z)$ be single-valued and regular for $|z| > R$. If one sets $z = \dfrac{1}{z'}$, then the function $\varphi(z')$ defined for $|z'| < \dfrac{1}{R}$ by $f(z) = f\left(\dfrac{1}{z'}\right) = \varphi(z')$ is single-valued and regular there with the possible exception (with respect to regularity) of the point $z' = 0$ itself. We now lay down the following definition.

Definition. *That behavior is assigned to the function $f(z)$ at infinity, which $\varphi(z')$ exhibits at $z' = 0$.*
  In detail:
  By our hypotheses, $\varphi(z')$ in $0 < |z'| < \dfrac{1}{R}$ admits of a Laurent expansion

$$(1) \qquad \varphi(z') = \sum_{n=-\infty}^{+\infty} b_n z'^n,$$

from which, according to the last paragraph, the behavior of $\varphi(z')$ at $z' = 0$ can be read off. This expansion differs only in notation from the Laurent expansion of $f(z)$ for $|z| > R$:

$$(2) \qquad f(z) = \sum_{n=-\infty}^{+\infty} a_n z^n,$$

which by hypothesis certainly exists; for, $a_n = -b_n$ and $z = \dfrac{1}{z'}$. Hence, if we carry over to $f(z)$ the behavior

of $\varphi(z')$ read off from (1), we see that "the point ∞" is now the isolated point in question, that the ascending part of (2) is to be considered as the principal part of $f(z)$, and that consequently
  a) $f(z)$ has an *essential singularity* at ∞ if an *infinite number* of *positive* powers appear in (2);
  b) $f(z)$ has a *pole of order β* at ∞ if only a *finite number*

of *positive* powers appear in (2), of which a_β is the last coefficient different from zero, $(\beta \geqq 1)$;

c) $f(z)$ is *regular* at ∞ if *no positive* powers appear in (2). In the last case, a_0 is taken to be the *value* of the function at ∞; i.e., $f(\infty) = a_0$. If $a_{-1} = \cdots = a_{-(p-1)} = 0$, $a_{-p} \neq 0$, then ∞ is an "a_0-point of order p."

Examples.

1. $\dfrac{1}{1-z}$ is regular at $z = \infty$, (because it is equal to $-\sum\limits_{n=1}^{\infty} \dfrac{1}{z^n}$ for $|z| > 1$), and has there a zero of the first order.

2. Every rational function for which the degree k of the denominator is greater than or equal to the degree m of the numerator is regular at $z = \infty$; $f(\infty)$ is zero or not zero according as $k > m$ or $k = m$, respectively.

3. Every rational function for which $k < m$ has a pole of order $m - k$ at $z = \infty$. In particular, a polynomial of degree m has a pole of order m at $z = \infty$.

4. e^z, $\sin z$, $\cos z$, and all other entire transcendental functions have an essential singularity at $z = \infty$.

Since we are only dealing with a transference of designation in these new definitions, the two theorems of the preceding paragraph are also valid for the point ∞ with suitable changes in wording.

Theorem 1. *If $f(z)$ has a pole at infinity, then, having chosen $G > 0$, one can always assign such a small neighborhood of ∞[1] that $|f(z)| > G$ for all points of that neighborhood (i.e., for all $|z| > R$, with R sufficiently large).*

And corresponding to the **Casorati-Weierstrass Theorem:**

Theorem 2. *If $f(z)$ has an essential singularity at ∞, then, having chosen the complex number c and the positive numbers ϵ and R, there always exist points z for which*

$$|z| > R \quad and \quad |f(z) - c| < \epsilon.$$

[1] A small "neighborhood of ∞" is understood to mean (see §2) the exterior of a large circle about the origin.

As an application of these considerations we prove the important **theorem of Riemann.**

Theorem 3. *If, in a certain neighborhood of a point z_0 (which may also be the point ∞), $f(z)$ is a single-valued and, apart from at z_0 itself, a regular function, then z_0 is*

a regular point, if and only if $f(z)$ is bounded in a neighborhood of z_0;

a pole, if and only if, having chosen $G > 0$, the neighborhood of z_0 can be contracted so that $|f(z)| > G$ everywhere in the resulting neighborhood;

an essential singularity, if and only if neither the first nor the second of the conditions just stated is satisfied.

Proof: By the hypotheses, $f(z)$ can be expanded in a Laurent series for the neighborhood of z_0. This series is of the form

$$f(z) = \sum_{n=-\infty}^{+\infty} a_n(z - z_0)^n \quad \text{or} \quad f(z) = \sum_{n=-\infty}^{+\infty} b_n\left(\frac{1}{z}\right)^n$$

according as z_0 lies in the finite part of the plane or is the point ∞, respectively.

The two theorems of this and the preceding paragraph, together with the fact that a function is bounded in a neighborhood of a regular point, show that the conditions stated are *necessary*. That they are *sufficient* follows immediately from the observation that the three possibilities for the behavior of $f(z)$ at z_0 are mutually exclusive and the only conceivable ones.

Exercise.   What kind of singularity does each of the functions

$$\frac{z^2 + 4}{e^z}, \cos z - \sin z, \cot z$$

have at the point $z = \infty$?

§33. The Residue Theorem

It $f(z)$ is regular in a neighborhood of z_0, then by Cauchy's theorem

$$\int f(z)dz = 0$$

if a small path C encircling the point z_0 in the positive sense is chosen as the path of integration. If, on the other hand, $f(z)$ has z_0 for an isolated singular point in whose neighborhood $f(z)$ is otherwise single-valued and regular, then the same integral will, in general, be different from zero. Its value can be found immediately. Since $f(z)$ can be expanded in a Laurent series for a neighborhood of z_0, $(0 < |z - z_0| < r)$, we have by §29 the relation

$$\frac{1}{2\pi i} \int_C f(z)dz = a_{-1}.$$

The value of this integral, or what is the same, the coefficient of that term of the Laurent expansion whose exponent is -1 is called the **residue** of $f(z)$ at z_0,[1] and the above formula represents in a certain sense an extension of Cauchy's theorem.

More generally, one can prove the following so-called **residue theorem.**

Theorem 1. *Let the function $f(z)$ be single-valued and regular in an arbitrary region $\mathfrak{G}$. If C is a simple closed path lying within $\mathfrak{G}$ and having only a finite number of singular points in its interior, then*

$$\frac{1}{2\pi i} \int_C f(z)dz = \begin{cases} \text{the sum of the residues of } f(z) \text{ at the singu-} \\ \text{lar points enclosed by } C. \end{cases}$$

Proof: If $z_1, z_2, \ldots, z_m$ are the finitely many singular points in question and if $C_1, C_2, \ldots, C_m$ are sufficiently small, positively-oriented circles about the respective centers $z_1, z_2, \ldots, z_m$, then by §14, Theorem 2

$$\frac{1}{2\pi i} \int_C f(z)dz$$

---

[1] z_0 is to be considered, once more, as lying in the finite part of the plane.

$$= \frac{1}{2\pi i} \int\limits_{C_1} f(z) \, dz + \frac{1}{2\pi i} \int\limits_{C_2} f(z) \, dz + \cdots + \frac{1}{2\pi i} \int\limits_{C_m} f(z) \, dz.$$

This proves the theorem, since the residues in question are the terms of the right member of this equation.

In applications the residue will, in general, be known from the Laurent expansion, so that it will be possible to determine the value of the integral. This residue theorem has numerous important applications, of which only a few chosen at random can be given here.

1. Under the hypotheses of the residue theorem, assume, for example, that $m = 0$, i.e., that $f(z)$ is regular in the whole interior of C, and, moreover, that $f(z) \neq 0$ along C. Then according to §21, Theorem 1, C can only enclose a finite number of zeros. Let these be the points $z_1, z_2, \ldots, z_m$ with the respective orders $\alpha_1, \alpha_2, \ldots, \alpha_m$. It is customary to consider a zero (or pole) of order α as an α-fold zero (or pole) and consequently count it α times in an enumeration. According to this, the number, N, of zeros of $f(z)$ in the interior of C is

$$N = \alpha_1 + \alpha_2 + \cdots + \alpha_m.$$

Theorem 2. *For this N we have*

$$N = \frac{1}{2\pi i} \int\limits_{C} \frac{f'(z)}{f(z)} \, dz.$$

Proof: The integrand is regular on the path C; $z_1, z_2, \ldots, z_m$ are singular points in the interior of C. It is readily seen that z_ν is a simple pole[1] with the residue α_ν. For, in general, if $f(z)$ has a zero of order α at ζ, then

$$f(z) = a_\alpha (z - \zeta)^\alpha + a_{\alpha+1} (z - \zeta)^{\alpha+1} + \cdots,$$
$$f'(z) = \alpha a_\alpha (z - \zeta)^{\alpha-1} + (\alpha + 1) a_{\alpha+1} (z - \zeta)^\alpha + \cdots.$$

[1] A pole of order unity.

Hence, since $a_\alpha \neq 0$,

$$\frac{f'(z)}{f(z)} = \frac{\alpha}{z - \zeta} + c_0 + c_1(z - \zeta) + \cdots$$

is the Laurent expansion of $\dfrac{f'(z)}{f(z)}$ valid for a certain neighborhood of ζ; the coefficients c_μ can easily be calculated from the coefficients a_ν. Therefore ζ is a simple pole with the residue α, as was asserted. It then follows immediately from the residue theorem that

$$\frac{1}{2\pi i} \int_C \frac{f'(z)}{f(z)} \, dz = \alpha_1 + \alpha_2 + \cdots + \alpha_m = N, \qquad \text{Q. E. D.}$$

2. If $f(z)$ has a pole of order β at ζ, one finds in exactly the same manner that $\dfrac{f'(z)}{f(z)}$ has a simple pole at ζ with the residue $-\beta$. Hence, if, in addition, the finitely many poles z_1', z_2', ..., z_k' with the respective orders β_1, β_2, ..., β_k lie within C, then

$$\frac{1}{2\pi i} \int_C \frac{f'(z)}{f(z)} \, dz = \alpha_1 + \alpha_2 + \cdots + \alpha_m - (\beta_1 + \beta_2 + \cdots + \beta_k).$$

Here $\beta_1 + \beta_2 + \cdots + \beta_k = P$ is the number of poles of $f(z)$ in the interior of C, in the same sense that N is the number of zeros there.   We have proved

Theorem 3.   *Let $f(z)$ be single-valued and regular in $\mathfrak{G}$, and let C be a simple closed path lying within $\mathfrak{G}$.   If $f(z) \neq 0$ along C, and if at most a finite number of singular points, all poles, lie in the interior of C, then*

$$\frac{1}{2\pi i} \int_C \frac{f'(z)}{f(z)} \, dz = N - P,$$

which is the number of zeros diminished by the number of poles of $f(z)$ in the interior of C, each point counted as often as its order requires.

3. The residue theorem furnishes a particularly important means for evaluating real definite integrals. We must be content with illustrating these applications by a very simple and transparent example.

As is readily found by indefinite integration,

$$(1) \qquad \int_{-\infty}^{+\infty} \frac{dx}{1 + x^2} = \pi.$$

With the aid of the residue theorem the integral can be evaluated as follows. Let C denote the path which extends from $z = -R$ rectilinearly to $+R$ and thence along the upper semicircle $|z| = R$ back to $-R$. Since

$$\frac{1}{1 + z^2} = \frac{1}{2i}\left(\frac{1}{z - i} - \frac{1}{z + i} \right),$$

this path encloses precisely one pole of $\dfrac{1}{1 + z^2}$ as soon as $R > 1$; at this pole the residue is $\dfrac{1}{2i}$. Consequently

$$\int_{C} \frac{dz}{1 + z^2} = 2\pi i \cdot \frac{1}{2i} = \pi.$$

Hence, also

$$(2) \qquad \int_{-R}^{+R} \frac{dx}{1 + x^2} + \int_{S} \frac{dz}{1 + z^2} = \pi,$$

if S denotes the aforementioned semicircle. By §11, Theorem 5 we have

$$\left| \int_{S} \frac{dz}{1 + z^2} \right| \leqq \frac{\pi R}{R^2 - 1},$$

and the right member tends to zero as $R \to +\infty$. If

we let $R \rightarrow + \infty$ in (2) we obtain equation (1) immediately.

In like manner one can evaluate the integral $\int\limits_{-\infty}^{+\infty} f(x)dx$ of every rational function $f(x)$ which is continuous for all real x and is such that the degree of its denominator exceeds that of the numerator by at least 2. It turns out that *the integral is equal to $2\pi i$ times the sum of the residues at the poles of $f(z)$ which lie in the upper half-plane.*

Exercises. 1. Let $f(z)$ have a zero of order α at z_1. What is the residue of

$$z \frac{f'(z)}{f(z)} \quad \text{and of} \quad \varphi(z) \frac{f'(z)}{f(z)}$$

at the point z_1 if $\varphi(z)$ denotes an arbitrary function which is regular at z_1? What is the answer if $f(z)$ has a pole of order β at z_1?

2. In connection with Exercise 1, evaluate and determine the meaning of

$$\frac{1}{2\pi i} \int\limits_C z \frac{f'(z)}{f(z)} \, dz \quad \text{and of} \quad \frac{1}{2\pi i} \int\limits_C \varphi(z) \frac{f'(z)}{f(z)} \, dz$$

if the hypotheses of Theorem 2 or of Theorem 3 of this paragraph are made with regard to $f(z)$ and C.

§34. Inverses of Analytic Functions

If a function $f(z)$ is regular at z_0 and if $f(z_0) = w_0$, then, because of the continuity of the function, the images of all points of a (sufficiently small) neighborhood of z_0 lie in a prescribed ϵ-neighborhood of w_0. Nothing follows from this as to whether a full neighborhood of w_0 is covered by these images or not, and whether, on the other hand, the image region can be covered more than once or not. In this respect we have the following theorem.

Theorem 1. *If $f(z)$ is regular in the circle K: $|z - z_0| < \rho$ and assumes the value $w_0 = f(z_0)$ to the first order at z_0, that is to say, $f'(z_0) \neq 0$, then a certain complete neighborhood of w_0 in the w-plane is covered precisely once by the image of a neighborhood of z_0.*

Proof: The function $f(z) - w_0$ is also regular in K. It has a simple zero[1] at z_0. Then according to §21, Theorem 2 it is possible to describe such a small circle K_1 with radius $\rho_1 < \rho$ about z_0 as center, that, except for z_0, there is no zero of $f(z) - w_0$ in its interior or on its boundary. $|f(z) - w_0|$ has a still positive minimum μ on the boundary of K_1. It can now be shown that every value w_1 which lies in the circle K' with radius μ and center w_0 in the w-plane is obtained for one and only one value $z = z_1$ in the interior of the circle K_1. That is to say, briefly, that $f(z) - w_1$ has precisely one zero, z_1, in the interior of K_1; or what is the same (by §33, Theorem 2), that the integral (containing the parameter w)[2]

$$(1) \qquad \frac{1}{2\pi i} \int_{K_1} \frac{f'(z)}{f(z) - w} dz$$

has the value unity if any particular point w_1 of K' is substituted for w. (For, $f(z) - w_1$ along K_1 is different from zero because of the meaning of μ.) On the basis of our hypotheses and §33, Theorem 2, the integral (1) certainly has the value unity for $w = w_0$ and must always be equal to a real integer, because of its meaning. Obviously it must always have the same value unity if we can show that its value represents a *continuous function* of w in K'. This follows, however, from

---

[1] A zero of order unity.
[2] $f'(z)$ in the numerator of the integrand is to be regarded as the derivative of the denominator with respect to z, with w constant.

the simple inequality

$$\left| \int_{K_1} \frac{f'(z)}{f(z) - w'} \, dz - \int_{K_1} \frac{f'(z)}{f(z) - w''} \, dz \right| \leqq |\, w'' - w' \,| \cdot \frac{M' \cdot l}{d^2}$$

in which M' denotes the maximum of $|\, f'(z) \,|$ along K_1, l the length of this path, and d the smaller of the distances of the points w', w'' from the boundary of the circle K'.

Thus, according to Theorem 1, for a given w in K', the point z in K_1 for which $f(z) = w$ is uniquely determined. By the requirements of the theorem, then, a single-valued function of w, $z = \varphi(w)$, is so defined in K' that always $f(\varphi(w)) = w$ or $\varphi(f(z)) = z$. The function $z = \varphi(w)$ is called the *inverse*[1] of the function $w = f(z)$, and we can express the content of Theorem 1 as follows:

For every function $f(z)$ which is regular at z_0 and for which $f'(z_0) \neq 0$ there exists a well-defined inverse function $z = \varphi(w)$ in a neighborhood of the point $w_0 = f(z_0)$.

With regard to this function we prove

Theorem 2. *The inverse function $z = \varphi(w)$ is a regular function of w in a neighborhood of w_0. For its derivative there we have (as in the real domain) the equation*

$$\varphi'(w) = \frac{1}{f'(z)} = \frac{1}{f'(\varphi(w))}.$$

The proof, which is almost self-evident, proceeds exactly as in the real domain. For fixed w_1 and neighboring w in K' we have

$$\frac{\varphi(w) - \varphi(w_1)}{w - w_1} = \frac{z - z_1}{f(z) - f(z_1)}.$$

Since distinct points z_1, z also correspond to distinct points w_1, w, respectively, and conversely, and since $z \to z_1$ as $w \to w_1$, one can read off the assertion from

---

[1] For this function, w is the independent and z the dependent variable.

this equality; $f'(z) \neq 0$ in a neighborhood of z_0 because $f'(z_0) \neq 0$.

Exercise. Show that a certain complete neighborhood of the point w_0 is covered precisely α times by the image of a neighborhood of z_0 if the value w_0 of the function $f(z)$ which is regular there is assumed to the order $\alpha (\geqq 1)$.

§35. Rational Functions

An analytic function, as we have already emphasized on p. 94, is but rarely obtainable in closed form. We have thus far met with this favorable case only in connection with the *entire* functions and the *rational* functions. If one wishes to undertake a classification of functions "purely function-theoretically," one must ignore entirely the *representation* of a function and only characterize it intrinsically (by its domain of values, the nature of its singular points, and the like). Thus, the entire functions, without any regard to the closed representation which is possible in this case, are characterized alone by the property of being regular in the entire plane. Theorems 2 and 5, §28 separate them "purely function-theoretically" into entire rational and entire transcendental functions.

The following two theorems characterize in a similar manner *the class of rational functions*.

Theorem 1. *A rational function has no singularities other than poles in the finite and infinite parts of the plane.*

The proof is contained in §31, Example 2 and §32, Examples 2 and 3; and we have already attained our goal when we prove the converse of this theorem.

Theorem 2. *If a single-valued function has no singularities other than poles in the finite part of the plane and at $z = \infty$, then it is a rational function.*

Proof: Since $f(z)$ is assumed to have at most a pole at $z = \infty$, it is regular everywhere outside a sufficiently large circle, i.e., in a certain "neighborhood of the point $z = \infty$," except possibly at $z = \infty$ itself. Hence, all singular points which may lie in the finite part of the plane lie within an assignable circle. Here there can only be a finite number of such points, because otherwise there would be a limit point of these singular points in this closed circle according to §3, Theorem 1. This point certainly would not be a pole, since a pole is necessarily isolated.

If there is no singular point in the finite part of the plane, then $f(z)$ is an *entire* function and in fact, according to §32, Example 4, an *entire rational* function (i.e., a *polynomial*). If, however, $z_1, z_2, \ldots, z_k$ are the finitely many singular points lying in the finite part of the plane, then $f(z)$ can be expanded in a neighborhood of each of them in a Laurent series which can contain only a finite number of negative powers:

$$f(z) = \sum_{n=0}^{\infty} a_n^{(\nu)} (z - z_\nu)^n + \frac{a_{-1}^{(\nu)}}{z - z_\nu} + \cdots + \frac{a_{-a_\nu}^{(\nu)}}{(z - z_\nu)^{\alpha_\nu}};$$

here α_ν denotes the order of the pole z_ν, ($\nu = 1, 2, \ldots, k$). If one denotes the principal part following the power series by $h_\nu(z)$, then $h_\nu(z)$ is a rational function which has the only singular point z_ν (pole of order α_ν) and is regular, and in fact equal to zero, at $z = \infty$.

The function

$$f(z) - h_1(z) - h_2(z) - \cdots - h_k(z)$$

is evidently an entire function, and indeed, since it too can only have at most a pole at infinity, a polynomial $g(z)$, which reduces to a constant (a polynomial of degree zero) if the point ∞ is a regular point.

Hence,

$$f(z) = g(z) + h_1(z) + h_2(z) + \cdots + h_k(z),[1]$$

which exhibits the rational character of $f(z)$.

---

[1] The terms $h_\nu(z)$ here are simply missing in the case that $f(z)$ is regular in the finite part of the plane; this case has already been treated.

Owing to the special form of the principal parts $h_\nu(z)$ we can also state the following theorem.

Theorem 3. *A rational function can be decomposed into partial fractions.* (Cf. *Elem.*, §40.)

We conclude with a second proof of the *fundamental theorem of algebra*, based on the residue theorem (cf. §28, 3 and *Elem.*, §39).

If $f(z)$ is a polynomial $a_0 + a_1 z + \cdots + a_m z^m$, ($m \geqq 1$, $a_m \neq 0$), then according to §28, Theorem 2 it is possible to describe a circle K with radius R about the origin as center such that $|f(z)| > 1$, and hence, that $f(z)$ has no zeros anywhere in its exterior or on its boundary. All existing zeros of $f(z)$ lie, then, in the interior of K.

Their number N, according to §33, Theorem 2, is:

$$N = \frac{1}{2\pi i} \int_K \frac{f'(z)}{f(z)} \, dz.$$

The Laurent expansion of the integrand, valid for $|z| > R$, begins with

$$\frac{m}{z} + \frac{c_2}{z^2} + \frac{c_3}{z^3} + \cdots,$$

where the coefficients c_ν need not be known. From this we can immediately read off the value of the integral as m, and hence

$$N = m;$$

i.e., a polynomial of degree m has precisely m zeros (roots) if each is counted as often as its order requires.

BIBLIOGRAPHY

The fundamental contributions of Cauchy, Riemann, and Weierstrass to the Theory of Functions are to be found in the collected works of these men:

Augustin Cauchy, *Œuvres Complètes*, Paris (Gauthier-Villars), 1882–1921.

Bernhard Riemann, *Gesammelte mathematische Werke*, 2d edition, Leipzig, 1892, Supplement 1902.

Karl Weierstrass, *Mathematische Werke*, Berlin, 1894–1927.

Some more recent comprehensive presentations are:

L. Bieberbach, *Lehrbuch der Funktionentheorie,*
  Vol. I: *Elemente der Funktionentheorie*, 4th edition, Leipzig, 1934,
  Vol. II: *Moderne Funktionentheorie*, 2d edition, Leipzig, 1931.

É. Borel, *Leçons sur la Théorie des Fonctions*, 3d edition, Paris, 1928.

H. Burkhardt, *Funktionentheoretische Vorlesungen*, vol. I, edited by G. Faber, Berlin, 1920–21.

E. T. Copson, *An Introduction to the Theory of Functions of a Complex Variable*, Oxford, 1935.

G. Doetsch, *Funktionentheorie.* (Constitutes ch. 15 of E. Pascal, *Repertorium der höheren Analysis*, vol. 1, part 2, 2d edition, Leipzig, 1927.)

É. Goursat, *Cours d'Analyse Mathématique*, vol. II, 5th edition, Paris, 1933.

——, *A Course in Mathematical Analysis*, vol. II, part I, translated by E. R. Hedrick and O. Dunkel, Boston, 1916.

J. Hadamard, *La Série de Taylor et son Prolongement Analytique*, (Coll. Scientia), 2d edition, edited by S. Mandelbrojt, Paris, 1926.

A. Hurwitz and R. Courant, *Funktionentheorie*, 3d edition, Berlin, 1929

C. Jordan, *Cours d'Analyse*, vol. I, 3d edition, Paris, 1909.

141

K. Knopp, *Theorie und Anwendung der unendlichen Reihen*, 3d edition, Berlin, 1931.

——, *Theory and Application of Infinite Series*, translated by R. C. Young from 2d German edition, London and Glasgow, 1928.

G. Kowalewski, *Die komplexen Veränderlichen und ihre Funktionen*, 2d edition, Leipzig, 1923.

H. v. Mangoldt and K. Knopp, *Einführung in die höhere Mathematik*, vols. II and III, Leipzig, 1932 and 1933.

W. F. Osgood, *Lehrbuch der Funktionentheorie*, vol. I, 5th edition, Leipzig, 1928.

É. Picard, *Traité d'Analyse*, vol. II, 3d edition, Paris, 1926.

E. C. Titchmarsh, *The Theory of Functions*, Oxford, 2d edition, 1939.

An additional reference is the

Enzyklopädie der mathematischen Wissenschaften, Leipzig, 1898 ff., of which parts II, 2 and II, 3 (Leipzig, 1901–1927) are devoted chiefly to the theory of functions.

INDEX

ORDINARY DIFFERENTIAL EQUATIONS
by E. L. Ince

The theory of ordinary differential equations in real and complex domains is here clearly explained and analyzed. The author covers not only classical theory, but also main developments of more recent times.

The pure mathematician will find valuable exhaustive sections on existence and nature of solutions, continuous transformation groups, the algebraic theory of linear differential systems, and the solution of differential equations by contour integration. The engineer and physicist will be interested in an especially fine treatment of the equations of Legendre, Bessel, and Mathieu; the transformations of Laplace and Mellin; the conditions for the oscillatory character of solutions of a differential equation; the relation between a linear differential system and an integral equation; the asymptotic development of characteristic numbers and functions; and many other topics.

PARTIAL CONTENTS: **Real Domain.** Elementary methods of integration. Existence and nature of solutions. Continuous transformation-groups. Linear differential equations — theory of, with constant coefficients, solutions of, algebraic theory of. Sturmian theory, its later developments. Boundary problems. **Complex Domain.** Existence theorems. Equations of first order. Non-linear equations of higher order. Solutions, systems, classifications of linear equations. Oscillation theorems.

"Will be welcomed by mathematicians, engineers, and others," MECH. ENGINEERING. " Highly recommended," ELECTRONICS INDUSTRIES. "Deserves the highest praise," BULLETIN, AM. MATH. SOC.

Historical appendix. Bibliography. Index. 18 figures. viii + 558pp. 5⅜ x 8.

S349 Paperbound **$2.45**

VECTOR AND TENSOR ANALYSIS
by G. E. Hay

First published in 1953, this is a simple clear introduction to classical vector and tensor analysis for students of engineering and mathematical physics. It is unusual for its appreciation of the problems which beset the beginning student, and its capable resolution of these problems.

Emphasis is upon vectors, with chapters discussing elementary vector operations, up to moments of vectors, linear vector differential equations; applications to plane, solid analytic and differential geometry; mechanics, with special reference to motion of a particle and of a system of particles; partial differentiation, with operator del and other operators; integration, with Green's theorem, Stokes's theory, irrotational and solenoidal vectors. Most important features of classical tensor analysis are also presented, with information on transformation of coordinates, contravariant and covariant tensors, metric tensors, conjugate tensors, geodesics, oriented Cartesian tensors, Christoffel symbols, and applications.

Many examples are worked in the text, while more than 200 problems are presented at the ends of chapters.

"Remarkably comprehensive, concise, and clear," INDUSTRIAL LABORATORIES. "A useful addition to the library on the subject," ELECTRONICS. "Considered as a condensed text in the classical manner, the book can well be recommended," NATURE (London).

66 figures. viii + 193pp. 5⅜ x 8.　　　　S109 Paperbound **$1.75**

LECTURES ON THE ICOSAHEDRON AND THE SOLUTION OF EQUATIONS OF THE FIFTH DEGREE by Felix Klein

This well-known monograph covers the solution of quintics in terms of the rotations of regular icosahedron around the axes of its symmetry. Still the only work unifying all previous knowledge on quintics, it is both an outstanding classic' of mathematics and an indispensable source book for those interested in higher algebra, geometry, or the mathematics of crystallography. An expert knowledge of higher mathematics is not required to follow the presentation, since considerable explanatory material is included.

Partial contents: THEORY OF THE ICOSAHEDRON ITSELF. Regular solids and theory of groups. Introduction of $(x + iy)$. Statement and discussion of the fundamental problem according to the theory of functions. On the algebraic character of our fundamental problem. General theorems and survey of the subject. THEORY OF EQUATIONS OF THE FIFTH DEGREE. Historical development of the theory of equations of the fifth degree. Introduction of geometrical material. The canonical equations of the fifth degree. The problem of the A's and the Jacobian equations of the 6th degree. The general equation of the fifth degree.

First edition in over 40 years of this outstanding classic which has brought up to $25 on the out of print market. 230 footnotes, mostly bibliographic. Second revised edition, with additional corrections. xvi + 289pp. 5⅜ x 8.

Paperbound, **$1.85**

INTRODUCTION TO THE THEORY OF FOURIER'S SERIES AND INTEGRALS

by H. S. Carslaw

As an introductory explanation of the theory of Fourier's series, this clear, detailed text has long been recognized as outstanding. This third revised edition contains tests for uniform convergence of series; a thorough treatment of term by term integration and the Second Theorem of Mean Value; enlarged sets of examples on Infinite Series and Integrals; and a section dealing with the Riemann-Lebesgue Theorem and its consequences. An appendix compares the Lebesgue Definite Integral with the classical Riemann Integral.

CONTENTS: Historical Introduction. Rational and irrational numbers. Infinite sequences and series. Functions of single variable. Limits and continuity. Definite integral. Theory of infinite series, whose terms are functions of a single variable. Definite integrals containing an arbitrary parameter. Fourier's series. Nature of convergence of Fourier's series and some properties of Fourier's constants. Approximation curves and Gibbs phenomenon in Fourier's series. Fourier's integrals. Appendices: Practical harmonic analysis and periodogram analysis; Lebesgue's theory of the definite integral.

"For the serious student of mathematical physics, anxious to have a firm grasp of Fourier theory as far as the Riemann integral will serve, Carslaw is still unsurpassed," MATHEMATICAL GAZETTE.

Bibliography. Index. 39 figures. 96 examples for students. xiii + 368pp. 5⅜ x 8.

S48 Paperbound **$1.95**

Catalogue of Dover
SCIENCE BOOKS

DIFFERENTIAL EQUATIONS
(ORDINARY AND PARTIAL DIFFERENTIAL)

INTRODUCTION TO THE DIFFERENTIAL EQUATIONS OF PHYSICS, L. Hopf. Especially valuable to engineer with no math beyond elementary calculus. Emphasizes intuitive rather than formal aspects of concepts. Partial contents: Law of causality, energy theorem, damped oscillations, coupling by friction, cylindrical and spherical coordinates, heat source, etc. 48 figures. 160pp. 5⅜ x 8. S120 Paperbound **$1.25**

INTRODUCTION TO BESSEL FUNCTIONS, F. Bowman. Rigorous, provides all necessary material during development, includes practical applications. Bessel functions of zero order, of any real order, definite integrals, asymptotic expansion, circular membranes, Bessel's solution to Kepler's problem, much more. "Clear . . . useful not only to students of physics and engineering, but to mathematical students in general," Nature. 226 problems. Short tables of Bessel functions. 27 figures. x + 135pp. 5⅜ x 8. S462 Paperbound **$1.35**

DIFFERENTIAL EQUATIONS, F. R. Moulton. Detailed, rigorous exposition of all non-elementary processes of solving ordinary differential equations. Chapters on practical problems; more advanced than problems usually given as illustrations. Includes analytic differential equations; variations of a parameter; integrals of differential equations; analytic implicit functions; problems of elliptic motion; sine-amplitude functions; deviation of formal bodies; Cauchy-Lipshitz process; linear differential equations with periodic coefficients; much more. Historical notes. 10 figures. 222 problems. xv + 395pp. 5⅜ x 8. S451 Paperbound **$2.00**

PARTIAL DIFFERENTIAL EQUATIONS OF MATHEMATICAL PHYSICS, A. G. Webster. Valuable sections on elasticity, compression theory, potential theory, theory of sound, heat conduction, wave propagation, vibration theory. Contents include: deduction of differential equations, vibrations, normal functions, Fourier's series. Cauchy's method, boundary problems, method of Riemann-Volterra, spherical, cylindrical, ellipsoidal harmonics, applications, etc. 97 figures. vii + 440pp. 5⅜ x 8. S263 Paperbound **$2.00**

ORDINARY DIFFERENTIAL EQUATIONS, E. L. Ince. A most compendious analysis in real and complex domains. Existence and nature of solutions, continuous transformation groups, solutions in an infinite form, definite integrals, algebraic theory. Sturmian theory, boundary problems, existence theorems, 1st order, higher order, etc. "Deserves highest praise, a notable addition to mathematical literature," Bulletin, Amer. Math. Soc. Historical appendix. 18 figures. viii + 558pp. 5⅜ x 8. S349 Paperbound **$2.55**

ASYMPTOTIC EXPANSIONS, A. Erdélyi. Only modern work available in English; unabridged reproduction of monograph prepared for Office of Naval Research. Discusses various procedures for asymptotic evaluation of integrals containing a large parameter; solutions of ordinary linear differential equations. vi + 108pp. 5⅜ x 8. S318 Paperbound **$1.35**

LECTURES ON CAUCHY'S PROBLEM, J. Hadamard. Based on lectures given at Columbia, Rome, discusses work of Riemann, Kirchhoff, Volterra, and author's own research on hyperbolic case in linear partial differential equations. Extends spherical cylindrical waves to apply to all (normal) hyperbolic equations. Partial contents: Cauchy's problem, fundamental formula, equations with odd number, with even number of independent variables; method of descent. 32 figures. iii + 316pp. 5⅜ x 8. S105 Paperbound **$1.75**

NUMBER THEORY

INTRODUCTION TO THE THEORY OF NUMBERS, L. E. Dickson. Thorough, comprehensive, with adequate coverage of classical literature. Not beyond beginners. Chapters on divisibility, congruences, quadratic residues and reciprocity, Diophantine equations, etc. Full treatment of binary quadratic forms without usual restriction to integral coefficients. Covers infinitude of primes, Fermat's theorem, Legendre's symbol, automorphs, Recent theorems of Thue, Siegal, much more. Much material not readily available elsewhere. 239 problems. 1 figure. viii + 183pp. 5⅜ x 8.   S342 Paperbound **$1.65**

ELEMENTS OF NUMBER THEORY, I. M. Vinogradov. Detailed 1st course for persons without advanced mathematics; 95% of this book can be understood by readers who have gone no farther than high school algebra. Partial contents: divisibility theory, important number theoretical functions, congruences, primitive roots and indices, etc. Solutions to problems, exercises. Tables of primes, indices, etc. Covers almost every essential formula in elementary number theory! "Welcome addition . . . reads smoothly," Bull. of the Amer. Math. Soc. 233 problems. 104 exercises. viii + 227pp. 5⅜ x 8.   S259 Paperbound **$1.60**

PROBABILITY THEORY AND INFORMATION THEORY

SELECTED PAPERS ON NOISE AND STOCHASTIC PROCESSES, edited by Prof. Nelson Wax, U. of Illinois. 6 basic papers for those whose work involves noise characteristics. Chandrasekhar, Uhlenbeck and Ornstein, Uhlenbeck and Ming, Rice, Doob. Included is Kac's Chauvenet-Prize winning "Random Walk." Extensive bibliography lists 200 articles, through 1953. 21 figures. 337pp. 6⅛ x 9¼.   S262 Paperbound **$2.35**

A PHILOSOPHICAL ESSAY ON PROBABILITIES, Marquis de Laplace. This famous essay explains without recourse to mathematics the principle of probability, and the application of probabilty to games of chance, natural philosophy, astronomy, many other fields. Translated from 6th French edition by F. W. Truscott, F. L. Emory. Intro. by E. T. Bell. 204pp. 5⅜ x 8.   S166 Paperbound **$1.25**

MATHEMATICAL FOUNDATIONS OF INFORMATION THEORY, A. I. Khinchin. For mathematicians, statisticians, physicists, cyberneticists, communications engineers, a complete, exact introduction to relatively new field. Entropy as a measure of a finite scheme, applications to coding theory, study of sources, channels and codes, detailed proofs of both Shannon theorems for any ergodic source and any stationary channel with finite memory, much more. "Presents for the first time rigorous proofs of certain fundamental theorems . . . quite complete . . . amazing expository ability," American Math. Monthly. vii + 120pp. 5⅜ x 8.   S434 Paperbound **$1.35**

VECTOR AND TENSOR ANALYSIS AND MATRIX THEORY

VECTOR AND TENSOR ANALYSIS, G. E. Hay. One of clearest introductions to increasingly important subject. Start with simple definitions, finish with sure mastery of oriented Cartesian vectors, Christoffel symbols, solenoidal tensors. Complete breakdown of plane, solid, analytical, differential geometry. Separate chapters on application. All fundamental formulae listed, demonstrated. 195 problems. 66 figures. viii + 193pp. 5⅜ x 8.   S109 Paperbound **$1.75**

APPLICATIONS OF TENSOR ANALYSIS, A. J. McConnell. Excellent text for applying tensor methods to such familiar subjects as dynamics, electricity, elasticity, hydrodynamics. Explains fundamental ideas and notation of tensor theory, geometrical treatment of tensor algebra, theory of differentiation of tensors, and a wealth of practical material. "The variety of fields treated and the presence of extremely numerous examples make this volume worth much more than its low price," Alluminio. Formerly titled "Applications of the Absolute Differential Calculus." 43 illustrations. 685 problems. xii + 381pp.   S373 Paperbound **$1.85**

VECTOR AND TENSOR ANALYSIS, A. P. Wills. Covers entire field, from dyads to non-Euclidean manifolds (especially detailed), absolute differentiation, the Riemann-Christoffel and Ricci-Einstein tensors, calculation of Gaussian curvature of a surface. Illustrations from electrical engineering, relativity theory, astro-physics, quantum mechanics. Presupposes only working knowledge of calculus. Intended for physicists, engineers, mathematicians. 44 diagrams. 114 problems. xxxii + 285pp. 5⅜ x 8.   S454 Paperbound **$1.75**

PHYSICS, ENGINEERING

MECHANICS, DYNAMICS, THERMODYNAMICS, ELASTICITY

MATHEMATICAL ANALYSIS OF ELECTRICAL AND OPTICAL WAVE-MOTION, H. Bateman. By one of century's most distinguished mathematical physicists, a practical introduction to developments of Maxwell's electromagnetic theory which directly concern the solution of partial differential equation of wave motion. Methods of solving wave-equation, polar-cylindrical coordinates, diffraction, transformation of coordinates, homogeneous solutions, electromagnetic fields with moving singularities, etc. 168pp. 5⅜ x 8.   S14 Paperbound **$1.60**

THERMODYNAMICS, Enrico Fermi. Unabridged reproduction of 1937 edition. Remarkable for clarity, organization; requires no knowledge of advanced math beyond calculus, only familiarity with fundamentals of thermometry, calorimetry. Partial Contents: Thermodynamic systems, 1st and 2nd laws, potentials; Entropy, phase rule; Reversible electric cells; Gaseous reactions: Van't Hoff reaction box, principle of LeChatelier; Thermodynamics of dilute solutions: osmotic, vapor pressures; boiling, freezing point; Entropy constant. 25 problems. 24 illustrations. x + 160pp. 5⅜ x 8.   S361 Paperbound **$1.75**

FOUNDATIONS OF POTENTIAL THEORY, O. D. Kellogg. Based on courses given at Harvard, suitable for both advanced and beginning mathematicians, Proofs rigorous, much material here not generally available elsewhere. Partial contents: gravity, fields of force, divergence theorem, properties of Newtonian potentials at points of free space, potentials as solutions of LaPlace's equation, harmonic functions, electrostatics, electric images, logarithmic potential, etc. ix + 384pp. 5⅜ x 8.   S144 Paperbound **$1.98**

DIALOGUES CONCERNING TWO NEW SCIENCES, Galileo Galilei. Classic of experimental science, mechanics, engineering, as enjoyable as it is important. Characterized by author as "superior to everything else of mine." Offers a lively exposition of dynamics, elasticity, sound, ballistics, strength of materials, scientific method. Translated by H. Grew, A. de Salvio. 126 diagrams. xxi + 288pp. 5⅜ x 8.   S99 Paperbound **$1.65**

THEORETICAL MECHANICS; AN INTRODUCTION TO MATHEMATICAL PHYSICS, J. S. Ames, F. D. Murnaghan. A mathematically rigorous development for advanced students, with constant practical applications. Used in hundreds of advanced courses. Unusually thorough coverage of gyroscopic baryscopic material, detailed analyses of Corilis acceleration, applications of Lagrange's equations, motion of double pendulum, Hamilton-Jacobi partial differential equations, group velocity, dispersion, etc. Special relativity included. 159 problems. 44 figures. ix + 462pp. 5⅜ x 8.   S461 Paperbound **$2.00**

STATICS AND THE DYNAMICS OF A PARTICLE, W. D. MacMillan. This is Part One of "Theoretical Mechanics." For over 3 decades a self-contained, extremely comprehensive advanced undergraduate text in mathematical physics, physics, astronomy, deeper foundations of engineering. Early sections require only a knowledge of geometry; later, a working knowledge of calculus. Hundreds of basic problems including projectiles to moon, harmonic motion, ballistics, transmission of power, stress and strain, elasticity, astronomical problems. 340 practice problems, many fully worked out examples. 200 figures. xvii + 430pp. 5⅜ x 8.
                    S467 Paperbound **$2.00**

THE THEORY OF THE POTENTIAL, W. D. MacMillan. This is Part Two of "Theoretical Mechanics." Comprehensive, well-balanced presentation, serving both as introduction and reference with regard to specific problems, for physicists and mathematicians. Assumes no prior knowledge of integral relations, all math is developed as needed. Includes: Attraction of Finite Bodies; Newtonian Potential Function; Vector Fields, Green and Gauss Theorems; Two-layer Surfaces; Spherical Harmonics; etc. "The great number of particular cases . . . should make the book valuable to geo-physicists and others actively engaged in practical applications of the potential theory," Review of Scientific Instruments. xii + 469pp. 5⅜ x 8.
                    S486 Paperbound **$2.25**

DYNAMICS OF A SYSTEM OF RIGID BODIES (Advanced Section), E. J. Routh. Revised 6th edition of a classic reference aid. Partial contents: moving axes, relative motion, oscillations about equilibrium, motion. Motion of a body under no forces, any forces. Nature of motion given by linear equations and conditions of stability. Free, forced vibrations, constants of integration, calculus of finite differences, variations, procession and mutation, motion of the moon, motion of string, chain, membranes. 64 figures. 498pp. 5⅜ x 8.
                    S229 Paperbound **$2.35**

THE DYNAMICS OF PARTICLES AND OF RIGID, ELASTIC, AND FLUID BODIES: BEING LECTURES ON MATHEMATICAL PHYSICS, A. G. Webster. Reissuing of classic fills need for comprehensive work on dynamics. Covers wide range in unusually great depth, applying ordinary, partial differential equations. Partial contents: laws of motion, methods applicable to systems of all sorts; oscillation, resonance, cyclic systems; dynamics of rigid bodies; potential theory; stress and strain; gyrostatics; wave, vortex motion; kinematics of a point; Lagrange's equations; Hamilton's principle; vectors; deformable bodies; much more not easily found together in one volume. Unabridged reprinting of 2nd edition. 20 pages on differential equations, higher analysis. 203 illustrations. xi + 588pp. 5⅜ x 8.   S522 Paperbound **$2.35**

PRINCIPLES OF MEC::ANICS, Heinrich Hertz. A classic of great interest in logic of science. Last work by great 19th century physicist, created new system of mechanics based upon space, time, mass; returns to axiomatic analysis, understanding of formal, structural aspects of science, taking into account logic, observation, a priori elements. Of great historical importance to Poincaré, Carnap, Einstein, Milne. 20 page introduction by R. S. Cohen, Wesleyan U., analyzes implications of Hertz's thought and logic of science. 13 page introduction by Helmholtz. xlii + 274pp. 5⅜ x 8.   S316 Clothbound **$3.50**
S317 Paperbound **$1.75**

MATHEMATICAL FOUNDATIONS OF STATISTICAL MECHANICS, A. I. Khinchin. A thoroughly up-to-date introduction, offering a precise and mathematically rigorous formulation of the problems of statistical mechanics. Provides analytical tools to replace many commonly used cumbersome concepts and devices. Partial contents: Geometry, kinematics of phase space; ergodic problem; theory of probability; central limit theorem; ideal monatomic gas; foundation of thermodynamics; dispersion, distribution of sum functions; etc. "Excellent introduction . . . clear, concise, rigorous," Quarterly of Applied Mathematics. viii + 179pp. 5⅜ x 8.   S146 Clothbound **$2.95**
S147 Paperbound **$1.35**

MECHANICS OF THE GYROSCOPE, THE DYNAMICS OF ROTATION, R. F. Deimel, Prof. of Mechanical Engineering, Stevens Inst. of Tech. Elementary, general treatment of dynamics of rotation, with special application of gyroscopic phenomena. No knowledge of vectors needed. Velocity of a moving curve, acceleration to a point, general equations of motion, gyroscopic horizon, free gyro, motion of discs, the damped gyro, 103 similar topics. Exercises. 75 figures. 208pp. 5⅜ x 8.   S66 Paperbound **$1.65**

MECHANICS VIA THE CALCULUS, P. W. Norris, W. S. Legge. Wide coverage, from linear motion to vector analysis; equations determining motion, linear methods, compounding of simple harmonic motions, Newton's laws of motion, Hooke's law, the simple pendulum, motion of a particle in 1 plane, centers of gravity, virtual work, friction, kinetic energy of rotating bodies, equilibrium of strings, hydrostatics, sheering stresses, elasticity, etc. Many worked-out examples. 550 problems. 3rd revised edition. xii + 367pp.   S207 Clothbound **$3.95**

A TREATISE ON THE MATHEMATICAL THEORY OF ELASTICITY, A. E. H. Love. An indispensable reference work for engineers, mathematicians, physicists, the most complete, authoritative treatment of classical elasticity in one volume. Proceeds from elementary notions of extension to types of strain, cubical dilatation, general theory of strains. Covers relation between mathematical theory of elasticity and technical mechanics; equilibrium of isotropic elastic solids and aelotropic solid bodies; nature of force transmission, Volterra's theory of dislocations; theory of elastic spheres in relation to tidal, rotational, gravitational effects on earth; general theory of bending; deformation of curved plates; buckling effects; much more. "The standard treatise on elasticity," American Math. Monthly. 4th revised edition. 76 figures. xviii + 643pp. 6⅛ x 9¼.   S174 Paperbound **$2.95**

NUCLEAR PHYSICS, QUANTUM THEORY, RELATIVITY

MESON PHYSICS, R. E. Marshak. Presents basic theory, and results of experiments with emphasis on theoretical significance. Phenomena involving mesons as virtual transitions avoided, eliminating some of least satisfactory predictions of meson theory. Includes production study of π mesons at nonrelativistic nucleon energies contracts between π and u mesons, phenomena associated with nuclear interaction of π mesons, etc. Presents early evidence for new classes of particles, indicates theoretical difficulties created by discovery of heavy mesons and hyperons. viii + 378pp. 5⅜ x 8.   S500 Paperbound **$1.95**

THE FUNDAMENTAL PRINCIPLES OF QUANTUM MECHANICS, WITH ELEMENTARY APPLICATIONS, E. C. Kemble. Inductive presentation, for graduate student, specialists in other branches of physics. Apparatus necessary beyond differential equations and advanced calculus developed as needed. Though general exposition of principles, hundreds of individual problems fully treated. "Excellent book . . . of great value to every student . . . rigorous and detailed mathematical discussion . .. has succeeded in keeping his presentation clear and understandable," Dr. Linus Pauling, J. of American Chemical Society. Appendices: calculus of variations, math. notes, etc. 611pp. 5⅝ x 8⅜.   T472 Paperbound **$2.95**

WAVE PROPAGATION IN PERIODIC STRUCTURES, L. Brillouin. General method, application to different problems: pure physics—scattering of X-rays in crystals, thermal vibration in crystal lattices, electronic motion in metals; problems in electrical engineering. Partial contents: elastic waves along 1-dimensional lattices of point masses. Propagation of waves along 1-dimensional lattices. Energy flow. 2, 3 dimensional lattices. Mathieu's equation. Matrices and propagation of waves along an electric line. Continuous electric lines. 131 illustrations. xii + 253pp. 5⅜ x 8.   S34 Paperbound **$1.85**

4

DOVER SCIENCE BOOKS

THEORY OF ELECTRONS AND ITS APPLICATION TO THE PHENOMENA OF LIGHT AND RADIANT HEAT, H. Lorentz. Lectures delivered at Columbia Univ., by Nobel laureate. Unabridged, form historical coverage of theory of free electrons, motion, absorption of heat, Zeeman effect, optical phenomena in moving bodies, etc. 109 pages notes explain more advanced sections. 9 figures. 352pp. 5⅜ x 8. S173 Paperbound **$1.85**

SELECTED PAPERS ON QUANTUM ELECTRODYNAMICS, edited by J. Schwinger. Facsimiles of papers which established quantum electrodynamics; beginning to present position as part of larger theory. First book publication in any language of collected papers of Bethe, Bloch, Dirac, Dyson, Fermi, Feynman, Heisenberg, Kusch, Lamb, Oppenheimer, Pauli, Schwinger, Tomonaga, Weisskopf, Wigner, etc. 34 papers: 29 in English, 1 in French, 3 in German, 1 in Italian. Historical commentary by editor. xvii + 423pp. 6⅛ x 9¼. S444 Paperbound **$2.45**

FOUNDATIONS OF NUCLEAR PHYSICS, edited by R. T. Beyer. 13 of the most important papers on nuclear physics reproduced in facsimile in the original languages; the papers most often cited in footnotes, bibliographies. Anderson, Curie, Joliot, Chadwick, Fermi, Lawrence, Cockroft, Hahn, Yukawa. Unparalleled bibliography: 122 double columned pages, over 4,000 articles, books, classified. 57 figures. 288pp. 6⅛ x 9¼. S19 Paperbound **$1.75**

THE THEORY OF GROUPS AND QUANTUM MECHANICS, H. Weyl. Schroedinger's wave equation, de Broglie's waves of a particle, Jordon-Hoelder theorem, Lie's continuous groups of transformations, Pauli exclusion principle, quantization of Mawell-Dirac field equations, etc. Unitary geometry, quantum theory, groups, application of groups to quantum mechanics, symmetry permutation group, algebra of symmetric transformations, etc. 2nd revised edition. xxii + 422pp. 5⅜ x 8. S268 Clothbound **$4.50**
S269 Paperbound **$1.95**

PHYSICAL PRINCIPLES OF THE QUANTUM THEORY, Werner Heisenberg. Nobel laureate discusses quantum theory; his own work, Compton, Schroedinger, Wilson, Einstein, many others. For physicists, chemists, not specialists in quantum theory. Only elementary formulae considered in text; mathematical appendix for specialists. Profound without sacrificing clarity. Translated by C. Eckart, F. Hoyt. 18 figures. 192pp. 5⅜ x 8.
S113 Paperbound **$1.25**

INVESTIGATIONS ON THE THEORY OF THE BROWNIAN MOVEMENT, Albert Einstein. Reprints from rare European journals, translated into English. 5 basic papers, including Elementary Theory of the Brownian Movement, written at request of Lorentz to provide a simple explanation. Translated by A. D. Cowper. Annotated, edited by R. Fürth. 33pp. of notes elucidate, give history of previous investigations. 62 footnotes. 124pp. 5⅜ x 8.
S304 Paperbound **$1.25**

THE PRINCIPLE OF RELATIVITY, E. Einstein, H. Lorentz, M. Minkowski, H. Weyl. The 11 basic papers that founded the general and special theories of relativity, translated into English. 2 papers by Lorentz on the Michelson experiment, electromagnetic phenomena. Minkowski's "Space and Time," and Weyl's "Gravitation and Electricity." 7 epoch-making papers by Einstein: "Electromagnetics of Moving Bodies," "Influence of Gravitation in Propagation of Light," "Cosmological Considerations," "General Theory," 3 others. 7 diagrams. Special notes by A. Sommerfeld. 224pp. 5⅜ x 8. S93 Paperbound **$1.75**

STATISTICS

ELEMENTARY STATISTICS, WITH APPLICATIONS IN MEDICINE AND THE BIOLOGICAL SCIENCES, F. E. Croxton. Based primarily on biological sciences, but can be used by anyone desiring introduction to statistics. Assumes no prior acquaintance, requires only modest knowledge of math. All basic formulas carefully explained, illustrated; all necessary reference tables included. From basic terms and concepts, proceeds to frequency distribution, linear, non-linear, multiple correlation, etc. Contains concrete examples from medicine, biology. 101 charts. 57 tables. 14 appendices. lv + 376pp. 5⅜ x 8. S506 Paperbound **$1.95**

ANALYSIS AND DESIGN OF EXPERIMENTS, H. B. Mann. Offers method for grasping analysis of variance, variance design quickly. Partial contents: Chi-square distribution, analysis of variance distribution, matrices, quadratic forms, likelihood ration tests, test of linear hypotheses, power of analysis, Galois fields, non-orthogonal data, interblock estimates, etc. 15pp. of useful tables. x + 195pp. 5 x 7⅜. S180 Paperbound **$1.45**

FREQUENCY CURVES AND CORRELATION, W. P. Elderton. 4th revised edition of standard work on classical statistics. Practical, one of few books constantly referred to for clear presentation of basic material. Partial contents: Frequency Distributions; Pearsons Frequency Curves; Theoretical Distributions; Standard Errors; Correlation Ratio—Contingency; Corrections for Moments, Beta, Gamma Functions; etc. Key to terms, symbols. 25 examples. 40 tables. 16 figures. xi + 272pp. 5½ x 8½. Clothbound **$1.49**

HYDRODYNAMICS, ETC.

HYDRODYNAMICS, Horace Lamb. Standard reference work on dynamics of liquids and gases. Fundamental theorems, equations, methods, solutions, background for classical hydrodynamics. Chapters: Equations of Motion, Integration of Equations in Special Gases, Vortex Motion, Tidal Waves, Rotating Masses of Liquids, etc. Excellently planned, arranged, Clear, lucid presentation. 6th enlarged, revised edition. Over 900 footnotes, mostly bibliographical. 119 figures. xv + 738pp. 6⅛ x 9¼. S256 Paperbound **$2.95**

HYDRODYNAMICS, A STUDY OF LOGIC, FACT, AND SIMILITUDE, Garrett Birkhoff. A stimulating application of pure mathematics to an applied problem. Emphasis is on correlation of theory and deduction with experiment. Examines recently discovered paradoxes, theory of modelling and dimensional analysis, paradox and error in flows and free boundary theory. Classical theory of virtual mass derived from homogenous spaces; group theory applied to fluid mechanics. 20 figures, 3 plates. xiii + 186pp. 5⅜ x 8. S22 Paperbound **$1.85**

HYDRODYNAMICS, H. Dryden, F. Murnaghan, H. Bateman. Published by National Research Council, 1932. Complete coverage of classical hydrodynamics, encyclopedic in quality. Partial contents: physics of fluids, motion, turbulent flow, compressible fluids, motion in 1, 2, 3 dimensions; laminar motion, resistance of motion through viscous fluid, eddy viscosity, discharge of gases, flow past obstacles, etc. Over 2900-item bibliography. 23 figures. 634pp. 5⅜ x 8. S303 Paperbound **$2.75**

ACOUSTICS AND OPTICS

PRINCIPLES OF PHYSICAL OPTICS, Ernst Mach. Classical examination of propagation of light, color, polarization, etc. Historical, philosophical treatment unequalled for breadth and readability. Contents: Rectilinear propagation, reflection, refraction, dioptrics, composition of light, periodicity, theory of interference, polarization, mathematical representation of properties, etc. 279 illustrations. 10 portraits. 324pp. 5⅜ x 8. S170 Paperbound **$1.75**

THE THEORY OF SOUND, Lord Rayleigh. Written by Nobel laureate, classical methods here will cover most vibrating systems likely to be encountered in practice. Complete coverage of experimental, mathematical aspects. Partial contents: Harmonic motions, lateral vibrations of bars, curved plates or shells, applications of Laplace's functions to acoustical problems, fluid friction, etc. First low-priced edition of this great reference-study work. Historical introduction by R. B. Lindsay. 1040pp. 97 figures. 5⅜ x 8. S292, S293, Two volume set, paperbound **$4.00**

THEORY OF VIBRATIONS, N. W. McLachlan. Based on exceptionally successful graduate course, Brown University. Discusses linear systems having 1 degree of freedom, forced vibrations of simple linear systems, vibration of flexible strings, transverse vibrations of bars and tubes, of circular plate, sound waves of finite amplitude, etc. 99 diagrams. 160pp. 5⅜ x 8. S190 Paperbound **$1.35**

APPLIED OPTICS AND OPTICAL DESIGN, A. E. Conrady. Thorough systematic presentation of physical and mathematical aspects, limited mostly to "real optics." Stresses practical problem of maximum aberration permissible without affecting performance. Ordinary ray tracing methods; complete theory ray tracing methods, primary aberrations; enough higher aberration to design telescopes, low powered microscopes, photographic equipment. Covers fundamental equations, extra-axial image points, transverse chromatic aberration, angular magnification, similar topics. Tables of functions of N. Over 150 diagrams. x + 518pp. 5⅜ x 8⅝. S366 Paperbound **$2.98**

RAYLEIGH'S PRINCIPLE AND ITS APPLICATIONS TO ENGINEERING, G. Temple, W. Bickley. Rayleigh's principle developed to provide upper, lower estimates of true value of fundamental period of vibrating system, or condition of stability of elastic system. Examples, rigorous proofs. Partial contents: Energy method of discussing vibrations, stability. Perturbation theory, whirling of uniform shafts. Proof, accuracy, successive approximations, applications of Rayleigh's theory. Numerical, graphical methods. Ritz's method. 22 figures. ix + 156pp. 5⅜ x 8. S307 Paperbound **$1.50**

OPTICKS, Sir Isaac Newton. In its discussion of light, reflection, color, refraction, theories of wave and corpuscular theories of light, this work is packed with scores of insights and discoveries. In its precise and practical discussions of construction of optical apparatus, contemporary understanding of phenomena, it is truly fascinating to modern scientists. Foreword by Albert Einstein. Preface by I. B. Cohen, Harvard. 7 pages of portraits, facsimile pages, letters, etc. cxvi + 414pp. 5⅜ x 8. S205 Paperbound **$2.00**

DOVER SCIENCE BOOKS

ON THE SENSATIONS OF TONE, Hermann Helmholtz. Using acoustical physics, physiology, experiment, history of music, covers entire gamut of musical tone: relation of music science to acoustics, physical vs. physiological acoustics, vibration, resonance, tonality, progression of parts, etc. 33 appendixes on various aspects of sound, physics, acoustics, music, etc. Translated by A. J. Ellis. New introduction by H. Margenau, Yale. 68 figures. 43 musical passages analyzed. Over 100 tables. xix + 576pp. 6⅛ x 9¼.
S114 Clothbound **$4.95**

ELECTROMAGNETICS, ENGINEERING, TECHNOLOGY

INTRODUCTION TO RELAXATION METHODS, F. S. Shaw. Describes almost all manipulative resources of value in solution of differential equations. Treatment is mathematical rather than physical. Extends general computational process to include almost all branches of applied math and physics. Approximate numerical methods are demonstrated, although high accuracy is obtainable without undue expenditure of time. 48pp. of tables for computing irregular star first and second derivatives, irregular star coefficients for second order equations, for fourth order equations. "Useful. . . . exposition is clear, simple . . . no previous acquaintance with numerical methods is assumed," Science Progress. 253 diagrams. 72 tables. 400pp. 5⅜ x 8.
S244 Paperbound **$2.45**

THE ELECTROMAGNETIC FIELD, M. Mason, W. Weaver. Used constantly by graduate engineers. Vector methods exclusively; detailed treatment of electrostatics, expansion methods, with tables converting any quantity into absolute electromagnetic, absolute electrostatic, practical units. Discrete charges, ponderable bodies. Maxwell field equations, etc. 416pp. 5⅜ x 8.
S185 Paperbound **$2.00**

ELASTICITY, PLASTICITY AND STRUCTURE OF MATTER, R. Houwink. Standard treatise on rheological aspects of different technically important solids: crystals, resins, textiles, rubber, clay, etc. Investigates general laws for deformations; determines divergences. Covers general physical and mathematical aspects of plasticity, elasticity, viscosity. Detailed examination of deformations, internal structure of matter in relation to elastic, plastic behaviour, formation of solid matter from a fluid, etc. Treats glass, asphalt, balata, proteins, baker's dough, others. 2nd revised, enlarged edition. Extensive revised bibliography in over 500 footnotes. 214 figures. xvii + 368pp. 6 x 9¼.
S385 Paperbound **$2.45**

DESIGN AND USE OF INSTRUMENTS AND ACCURATE MECHANISM, T. N. Whitehead. For the instrument designer, engineer; how to combine necessary mathematical abstractions with independent observations of actual facts. Partial contents: instruments and their parts, theory of errors, systematic errors, probability, short period errors, erratic errors, design precision, kinematic, semikinematic design, stiffness, planning of an instrument, human factor, etc. 85 photos, diagrams. xii + 288pp. 5⅜ x 8.
S270 Paperbound **$1.95**

APPLIED HYDRO- AND AEROMECHANICS, L. Prandtl, O. G. Tietjens. Presents, for most part, methods valuable to engineers. Flow in pipes, boundary layers, airfoil theory, entry conditions, turbulent flow, boundary layer, determining drag from pressure and velocity, etc. "Will be welcomed by all students of aerodynamics," Nature. Unabridged, unaltered. An Engineering Society Monograph, 1934. Index. 226 figures. 28 photographic plates illustrating flow patterns. xvi + 311pp. 5⅜ x 8.
S375 Paperbound **$1.85**

FUNDAMENTALS OF HYDRO- AND AEROMECHANICS, L. Prandtl, O. G. Tietjens. Standard work, based on Prandtl's lectures at Goettingen. Wherever possible hydrodynamics theory is referred to practical considerations in hydraulics, unifying theory and experience. Presentation extremely clear. Though primarily physical, proofs are rigorous and use vector analysis to a great extent. An Engineering Society Monograph, 1934. "Still recommended as an excellent introduction to this area," Physikalische Blätter. 186 figures. xvi + 270pp. 5⅜ x 8.
S374 Paperbound **$1.85**

GASEOUS CONDUCTORS: THEORY AND ENGINEERING APPLICATIONS, J. D. Cobine. Indispensable text, reference, to gaseous conduction phenomena, with engineering viewpoint prevailing throughout. Studies kinetic theory of gases, ionization, emission phenomena; gas breakdown, spark characteristics, glow, discharges; engineering applications in circuit interrupters, rectifiers, etc. Detailed treatment of high pressure arcs (Suits); low pressure arcs (Langmuir, Tonks). Much more. "Well organized, clear, straightforward," Tonks, Review of Scientific Instruments. 83 practice problems. Over 600 figures. 58 tables. xx + 606pp. 5⅜ x 8.
S442 Paperbound **$2.75**

PHOTOELASTICITY: PRINCIPLES AND METHODS, H. T. Jessop, F. C. Harris. For engineer, specific problems of stress analysis. Latest time-saving methods of checking calculations in 2-dimensional design problems, new techniques for stresses in 3 dimensions, lucid description of optical systems used in practical photoelectricity. Useful suggestions, hints based on on-the-job experience included. Partial contents: strain, stress-strain relations, circular disc under thrust along diameter, rectangular block with square hold under vertical thrust, simply supported rectangular beam under central concentrated load, etc. Theory held to minimum, no advanced mathematical training needed. 164 illustrations. viii + 184pp. 6⅛ x 9¼.
S137 Clothbound **$3.75**

MICROWAVE TRANSMISSION DESIGN DATA, T. Moreno. Originally classified, now rewritten, enlarged (14 new chapters) under auspices of Sperry Corp. Of immediate value or reference use to radio engineers, systems designers, applied physicists, etc. Ordinary transmission line theory; attenuation; parameters of coaxial lines; flexible cables; tuneable wave guide impedance transformers; effects of temperature, humidity; much more. "Packed with information . . . theoretical discussions are directly related to practical questions," U. of Royal Naval Scientific Service. Tables of d electrics, flexible cable, etc. ix + 248pp. 5⅜ x 8.
S549 Paperbound **$1.50**

THE THEORY OF THE PROPERTIES OF METALS AND ALLOYS, H. F. Mott, H. Jones. Quantum methods develop mathematical models showing interrelationship of fundamental chemical phenomena wtih crystal structure, electrical, optical properties, etc. Examines electron motion in applied field, cohesion, heat capacity, refraction, noble metals, transition and di-valent metals, etc. "Exposition is as clear . . . mathematical treatment as simple and reliable as we have become used to expect of . . . Prof. Mott," Nature. 138 figures. xiii + 320pp. 5⅜ x 8.
S456 Paperbound **$1.85**

THE MEASUREMENT OF POWER SPECTRA FROM THE POINT OF VIEW OF COMMUNICATIONS ENGINEERING, R. B. Blackman, J. W. Tukey. Pathfinding work reprinted from "Bell System Technical Journal." Various ways of getting practically useful answers in power spectra measurement, using results from both transmission and statistical estimation theory. Treats: Autocovariance, Functions and Power Spectra, Distortion, Heterodyne Filtering, Smoothing, Decimation Procedures, Transversal Filtering, much more. Appendix reviews fundamental Fourier techniques. Index of notation. Glossary of terms. 24 figures. 12 tables. 192pp. 5⅝ x 8⅝.
S507 Paperbound **$1.85**

TREATISE ON ELECTRICITY AND MAGNETISM, James Clerk Maxwell. For more than 80 years a seemingly inexhaustible source of leads for physicists, mathematicians, engineers. Total of 1082pp. on such topics as Measurement of Quantities, Electrostatics, Elementary Mathematical Theory of Electricity, Electrical Work and Energy in a System of Conductors, General Theorems, Theory of Electrical Images, Electrolysis, Conduction, Polarization, Dielectrics, Resistance, much more. "The greatest mathematical physicist since Newton," Sir James Jeans. 3rd edition. 107 figures, 21 plates. 1082pp. 5⅜ x 8.
S186 Clothbound **$4.95**

CHEMISTRY AND PHYSICAL CHEMISTRY

THE PHASE RULE AND ITS APPLICATIONS, Alexander Findlay. Covers chemical phenomena of 1 to 4 multiple component systems, the "standard work on the subject" (Nature). Completely revised, brought up to date by A. N. Campbell, N. O. Smith. New material on binary, tertiary liquid equilibria, solid solutions in ternary systems, quinary systems of salts, water, etc. Completely revised to triangular coordinates in ternary systems, clarified graphic representation, solid models, etc. 9th revised edition. 236 figures. 505 footnotes, mostly bibliographic. xii + 449pp. 5⅜ x 8.
S92 Paperbound **$2.45**

DYNAMICAL THEORY OF GASES, James Jeans. Divided into mathematical, physical chapters for convenience of those not expert in mathematics. Discusses mathematical theory of gas in steady state, thermodynamics, Bolzmann, Maxwell, kinetic theory, quantum theory, exponentials, etc. "One of the classics of scientific writing . . . as lucid and comprehensive an exposition of the kinetic theory as has ever been written," J. of Institute of Engineers. 4th enlarged edition, with new material on quantum theory, quantum dynamics, etc. 28 figures. 444pp. 6⅛ x 9¼.
S136 Paperbound **$2.45**

POLAR MOLECULES, Pieter Debye. Nobel laureate offers complete guide to fundamental electrostatic field relations, polarizability, molecular structure. Partial contents: electric intensity, displacement, force, polarization by orientation, molar polarization, molar refraction, halogen-hydrides, polar liquids, ionic saturation, dielectric constant, etc. Special chapter considers quantum theory. "Clear and concise . . . coordination of experimental results with theory will be readily appreciated," Electronics Industries. 172pp. 5⅜ x 8.
S63 Clothbound **$3.50**
S64 Paperbound **$1.50**

ATOMIC SPECTRA AND ATOMIC STRUCTURE, G. Herzberg. Excellent general survey for chemists, physicists specializing in other fields. Partial contents: simplest line spectra, elements of atomic theory; multiple structure of line spectra, electron spin; building-up principle, periodic system of elements; finer details of atomic spectra; hyperfine structure of spectral lines; some experimental results and applications. 80 figures. 20 tables. xiii + 257pp. 5⅜ x 8.
S115 Paperbound **$1.95**

TREATISE ON THERMODYNAMICS, Max Planck. Classic based on his original papers. Brilliant concepts of Nobel laureate make no assumptions regarding nature of heat, rejects earlier approaches of Helmholtz, Maxwell, to offer uniform point of view for entire field. Seminal work by founder of quantum theory, deducing new physical, chemical laws. A standard text, an excellent introduction to field for students with knowledge of elementary chemistry, physics, calculus. 3rd English edition. xvi + 297pp. 5⅜ x 8.
S219 Paperbound **$1.75**

DOVER SCIENCE BOOKS

KINETIC THEORY OF LIQUIDS, J. Frenkel. Regards kinetic theory of liquids as generalization, extension of theory of solid bodies, covers all types of arrangements of solids; thermal displacements of atoms; interstitial atoms, ions; orientational, rotational motion of molecules; transition between states of matter. Mathematical theory developed close to physical subject matter. "Discussed in a simple yet deeply penetrating fashion . . . will serve as seeds for a great many basic and applied developments in chemistry," J. of the Amer. Chemical Soc. 216 bibliographical footnotes. 55 figures. xi + 485pp. 5⅜ x 8.
S94 Clothbound **$3.95**
S95 Paperbound **$2.45**

ASTRONOMY

OUT OF THE SKY, H. H. Nininger. Non-technical, comprehensive introduction to "meteoritics" —science concerned with arrival of matter from outer space. By one of world's experts on meteorites, this book defines meteors and meteorites; studies fireball clusters and processions, meteorite composition, size, distribution, showers, explosions, origins, much more. viii + 336pp. 5⅜ x 8.
T519 Paperbound **$1.85**

AN INTRODUCTION TO THE STUDY OF STELLAR STRUCTURE, S. Chandrasekhar. Outstanding treatise on stellar dynamics by one of greatest astro-physicists. Examines relationship between loss of energy, mass, and radius of stars in steady state. Discusses thermodynamic laws from Caratheodory's axiomatic standpoint; adiabatic, polytropic laws; work of Ritter, Emden, Kelvin, etc.; Stroemgren envelopes as starter for theory of gaseous stars; Gibbs statistical mechanics (quantum); degenerate stellar configuration, theory of white dwarfs; etc. "Highest level of scientific merit," Bulletin. Amer. Math. Soc. 33 figures. 509pp. 5⅜ x 8.
S413 Paperbound **$2.75**

LES MÉTHODES NOVELLES DE LA MÉCANIQUE CÉLESTE, H. Poincaré. Complete French text of one of Poincaré's most important works. Revolutionized celestial mechanics: first use of integral invariants, first major application of linear differential equations, study of periodic orbits, lunar motion and Jupiter's satellites, three body problem, and many other important topics. "Started a new era . . . so extremely modern that even today few have mastered his weapons," E. T. Bell. 3 volumes. Total 1282pp. 6⅛ x 9¼.
Vol. 1 S401 Paperbound **$2.75**
Vol. 2 S402 Paperbound **$2.75**
Vol. 3 S403 Paperbound **$2.75**
The set **$7.50**

THE REALM OF THE NEBULAE, E. Hubble. One of the great astronomers of our time presents his concept of "island universes," and describes its effect on astronomy. Covers velocity-distance relation; classification, nature, distances, general field of nebulae; cosmological theories; nebulae in the neighborhood of the Milky way; etc. 39 photos, including velocity-distance relations shown by spectrum comparison. "One of the most progressive lines of astronomical research," The Times, London. New Introduction by A. Sandage. 55 illustrations. xxiv + 201pp. 5⅜ x 8.
S455 Paperbound **$1.50**

HOW TO MAKE A TELESCOPE, Jean Texereau. Design, build an f/6 or f/8 Newtonian type reflecting telescope, with altazimuth Couder mounting, suitable for planetary, lunar, and stellar observation. Covers every operation step-by-step, every piece of equipment. Discusses basic principles of geometric and physical optics (unnecessary to construction), comparative merits of reflectors, refractors. A thorough discussion of eyepieces, finders, grinding, installation, testing, etc. 241 figures, 38 photos, show almost every operation and tool. Potential errors are anticipated. Foreword by A. Couder. Sources of supply. xiii + 191pp. 6¼ x 10.
T464 Clothbound **$3.50**

BIOLOGICAL SCIENCES

THE BIOLOGY OF THE AMPHIBIA, G. K. Noble, Late Curator of Herpetology at Am. Mus. of Nat. Hist. Probably most used text on amphibia, most comprehensive, clear, detailed. 19 chapters, 85 page supplement: development; heredity; life history; speciation; adaptation; sex, integument, respiratory, circulatory, digestive, muscular, nervous systems; instinct, intelligence, habits, economic value classification, environment relationships, etc. "Nothing comparable to it," C. H. Pope, curator of Amphibia, Chicago Mus. of Nat. Hist. 1047 item bibliography. 174 illustrations. 600pp. 5⅜ x 8.
S206 Paperbound **$2.98**

THE ORIGIN OF LIFE, A. I. Oparin. A classic of biology. This is the first modern statement of theory of gradual evolution of life from nitrocarbon compounds. A brand-new evaluation of Oparin's theory in light of later research, by Dr. S. Margulis, University of Nebraska. xxv + 270pp. 5⅜ x 8.
S213 Paperbound **$1.75**

THE BIOLOGY OF THE LABORATORY MOUSE, edited by G. D. Snell. Prepared in 1941 by staff of Roscoe B. Jackson Memorial Laboratory, still the standard treatise on the mouse, assembling enormous amount of material for which otherwise you spend hours of research. Embryology, reproduction, histology, spontaneous neoplasms, gene and chromosomes mutations, genetics of spontaneous tumor formations, of tumor transplantation, endocrine secretion and tumor formation, milk influence and tumor formation, inbred, hybrid animals, parasites, infectious diseases, care and recording. "A wealth of information of vital concern. . . . recommended to all who could use a book on such a subject," Nature. Classified bibliography of 1122 items. 172 figures, including 128 photos. ix + 497pp. 6⅛ x 9¼.
S248 Clothbound **$6.00**

THE TRAVELS OF WILLIAM BARTRAM, edited by Mark Van Doren. Famous source-book of American anthropology, natural history, geography, is record kept by Bartram in 1770's on travels through wilderness of Florida, Georgia, Carolinas. Containing accurate, beautiful descriptions of Indians, settlers, fauna, flora, it is one of finest pieces of Americana ever written. 13 original illustrations. 448pp. 5⅜ x 8.
T13 Paperbound **$2.00**

BEHAVIOUR AND SOCIAL LIFE OF THE HONEYBEE, Ronald Ribbands. Outstanding scientific study; a compendium of practically everything known of social life of honeybee. Stresses behaviour of individual bees in field, hive. Extends von Frisch's experiments on communication among bees. Covers perception of temperature, gravity, distance, vibration; sound production; glands; structural differences; wax production; temperature regulation; recognition, communication; drifting, mating behaviour, other highly interesting topics. "This valuable work is sure of a cordial reception by laymen, beekeepers and scientists," Prof. Karl von Frisch, Brit. J. of Animal Behaviour. Bibliography of 690 references. 127 diagrams, graphs, sections of bee anatomy, fine photographs. 352pp.
S410 Clothbound **$4.50**

ELEMENTS OF MATHEMATICAL BIOLOGY, A. J. Lotka. Pioneer classic, 1st major attempt to apply modern mathematical techniques on large scale to phenomena of biology, biochemistry, psychology, ecology, similar life sciences. Partial contents: Statistical meaning of irreversibility; Evolution as redistribution; Equations of kinetics of evolving systems; Chemical, inter-species equilibrium; parameters of state; Energy transformers of nature, etc. Can be read with profit by even those having no advanced math; unsurpassed as study-reference. Formerly titled "Elements of Physical Biology." 72 figures. xxx + 460pp. 5⅜ x 8.
S346 Paperbound **$2.45**

TREES OF THE EASTERN AND CENTRAL UNITED STATES AND CANADA, W. M. Harlow. Serious middle-level text covering more than 140 native trees, important escapes, with information on general appearance, growth habit, leaf forms, flowers, fruit, bark, commercial use, distribution, habitat, woodlore, etc. Keys within text enable you to locate various species easily, to know which have edible fruit, much more useful, interesting information. "Well illustrated to make identification very easy," Standard Cat. for Public Libraries. Over 600 photographs, figures. xiii + 288pp. 5⅝ x 6½.
T395 Paperbound **$1.35**

FRUIT KEY AND TWIG KEY TO TREES AND SHRUBS (Fruit key to Northeastern Trees, Twig key to Deciduous Woody Plants of Eastern North America), W. M. Harlow. Only guides with photographs of every twig, fruit described. Especially valuable to novice. Fruit key (both deciduous trees, evergreens) has introduction on seeding, organs involved, types, habits. Twig key introduction treats growth, morphology. In keys proper, identification is almost automatic. Exceptional work, widely used in university courses, especially useful for identification in winter, or from fruit or seed only. Over 350 photos, up to 3 times natural size. Index of common, scientific names, in each key. xvii + 125pp. 5⅝ x 8⅜.
T511 Paperbound **$1.25**

INSECT LIFE AND INSECT NATURAL HISTORY, S. W. Frost. Unusual for emphasizing habits, social life, ecological relations of insects rather than more academic aspects of classification, morphology. Prof. Frost's enthusiasm and knowledge are everywhere evident as he discusses insect associations, specialized habits like leaf-rolling, leaf mining, case-making, the gall insects, boring insects, etc. Examines matters not usually covered in general works: insects as human food; insect music, musicians; insect response to radio waves; use of insects in art, literature. "Distinctly different, possesses an individuality all its own," Journal of Forestry. Over 700 illustrations. Extensive bibliography. x + 524pp. 5⅜ x 8.
T519 Paperbound **$2.49**

A WAY OF LIFE, AND OTHER SELECTED WRITINGS, Sir William Osler. Physician, humanist, Osler discusses brilliantly Thomas Browne, Gui Patin, Robert Burton, Michael Servetus, William Beaumont, Laennec. Includes such favorite writing as title essay, "The Old Humanities and the New Science," "Books and Men," "The Student Life," 6 more of his best discussions of philosophy, literature, religion. "The sweep of his mind and interests embraced every phase of human activity," G. L. Keynes. 5 photographs. Introduction by G. L. Keynes, M.D., F.R.C.S. xx + 278pp. 5⅜ x 8.
T488 Paperbound **$1.50**

THE GENETICAL THEORY OF NATURAL SELECTION, R. A. Fisher. 2nd revised edition of vital reviewing of Darwin's Selection Theory in terms of particulate inheritance, by one of greatest authorities on experimental, theoretical genetics. Theory stated in mathematical form. Special features of particulate inheritance are examined: evolution of dominance, maintenance of specific variability, mimicry, sexual selection, etc. 5 chapters on man's special circumstances as a social animal. 16 photographs. x + 310pp. 5⅜ x 8.
S466 Paperbound **$1.85**

THE AUTOBIOGRAPHY OF CHARLES DARWIN, AND SELECTED LETTERS, edited by Francis Darwin. Darwin's own record of early life; historic voyage aboard "Beagle;" furore surrounding evolution, his replies; reminiscences of his son. Letters to Henslow, Lyell, Hooker, Huxley, Wallace, Kingsley, etc., and thoughts on religion, vivisection. We see how he revolutionized geology with concepts of ocean subsidence; how his great books on variation of plants and animals, primitive man, expression of emotion among primates, plant fertilization, carnivorous plants, protective coloration, etc., came into being. 365pp. 5⅜ x 8.
T479 Paperbound **$1.65**

ANIMALS IN MOTION, Eadweard Muybridge. Largest, most comprehensive selection of Muybridge's famous action photos of animals, from his "Animal Locomotion." 3919 high-speed shots of 34 different animals, birds, in 123 types of action; horses, mules, oxen, pigs, goats, camels, elephants, dogs, cats guanacos, sloths, lions, tigers, jaguars, raccoons, baboons, deer, elk, gnus, kangaroos, many others, walking, running, flying, leaping. Horse alone in over 40 ways. Photos taken against ruled backgrounds; most actions taken from 3 angles at once: 90°, 60°, rear. Most plates original size. Of considerable interest to scientists as biology classic, records of actual facts of natural history, physiology. "Really marvelous series of plates," Nature. "Monumental work," Waldemar Kaempffert. Edited by L. S. Brown, 74 page introduction on mechanics of motion. 340pp. of plates. 3919 photographs. 416pp. Deluxe binding, paper. (Weight: 4½ lbs.) 7⅛ x 10⅝.
T203 Clothbound **$10.00**

THE HUMAN FIGURE IN MOTION, Eadweard Muybridge. New edition of great classic in history of science and photography, largest selection ever made from original Muybridge photos of human action: 4789 photographs, illustrating 163 types of motion: walking, running, lifting, etc. in time-exposure sequence photos at speeds up to 1/6000th of a second. Men, women, children, mostly undraped, showing bone, muscle positions against ruled backgrounds, mostly taken at 3 angles at once. Not only was this a great work of photography, acclaimed by contemporary critics as work of genius, but it was also a great 19th century landmark in biological research. Historical introduction by Prof. Robert Taft, U. of Kansas. Plates original size, full of detail. Over 500 action strips. 407pp. 7¾ x 10⅝. Deluxe edition.
7204 Clothbound **$10.00**

AN INTRODUCTION TO THE STUDY OF EXPERIMENTAL MEDICINE, Claude Bernard. 90-year old classic of medical science, only major work of Bernard available in English, records his efforts to transform physiology into exact science. Principles of scientific research illustrated by specified case histories from his work; roles of chance, error, preliminary false conclusion, in leading eventually to scientific truth; use of hypothesis. Much of modern application of mathematics to biology rests on foundation set down here. "The presentation is polished . . . reading is easy," Revue des questions scientifiques. New foreword by Prof. I. B. Cohen, Harvard U. xxv + 266pp. 5⅜ x 8.
T400 Paperbound **$1.50**

STUDIES ON THE STRUCTURE AND DEVELOPMENT OF VERTEBRATES, E. S. Goodrich. Definitive study by greatest modern comparative anatomist. Exhaustive morphological, phylogenetic expositions of skeleton, fins, limbs, skeletal visceral arches, labial cartilages, visceral clefts, gills, vascular, respiratory, excretory, periphal nervous systems, etc., from fish to higher mammals. "For many a day this will certainly be the standard textbook on Vertebrate Morphology in the English language," Journal of Anatomy. 754 illustrations. 69 page biographical study by C. C. Hardy. Bibliography of 1186 references. Two volumes, total 906pp. 5⅜ x 8.
Two vol. set S449, 450 Paperbound **$5.00**

EARTH SCIENCES

THE EVOLUTION OF IGNEOUS BOOKS, N. L. Bowen. Invaluable serious introduction applies techniques of physics, chemistry to explain igneous rock diversity in terms of chemical composition, fractional crystallization. Discusses liquid immiscibility in silicate magmas, crystal sorting, liquid lines of descent, fractional resorption of complex minerals, petrogen, etc. Of prime importance to geologists, mining engineers; physicists, chemists working with high temperature, pressures. "Most important," Times, London. 263 bibliographic notes. 82 figures. xviii + 334pp. 5⅜ x 8.
S311 Paperbound **$1.85**

GEOGRAPHICAL ESSAYS, M. Davis. Modern geography, geomorphology rest on fundamental work of this scientist. 26 famous essays present most important theories, field researches. Partial contents: Geographical Cycle; Plains of Marine, Subaerial Denudation; The Peneplain; Rivers, Valleys of Pennsylvania; Outline of Cape Cod; Sculpture of Mountains by Glaciers; etc. "Long the leader and guide," Economic Geography. "Part of the very texture of geography . . . models of clear thought," Geographic Review. 130 figures. vi + 777pp. 5⅜ x 8.
S383 Paperbound **$2.95**

URANIUM PROSPECTING, H. L. Barnes. For immediate practical use, professional geologist considers uranium ores, geological occurrences, field conditions, all aspects of highly profitable occupation. "Helpful information . . . easy-to-use, easy-to-find style," Geotimes. x + 117pp. 5⅜ x 8.
T309 Paperbound **$1.00**

DE RE METALLICA, Georgius Agricola. 400 year old classic translated, annotated by former President Herbert Hoover. 1st scientific study of mineralogy, mining, for over 200 years after its appearance in 1556 the standard treatise. 12 books, exhaustively annotated, discuss history of mining, selection of sites, types of deposits, making pits, shafts, ventilating, pumps, crushing machinery; assaying, smelting, refining metals; also salt alum, nitre, glass making. Definitive edition, with all 289 16th century woodcuts of original. Biographical, historical introductions. Bibliography, survey of ancient authors. Indexes. A fascinating book for anyone interested in art, history of science, geology, etc. Deluxe Edition. 289 illustrations. 672pp. 6¾ x 10. Library cloth. S6 Clothbound **$10.00**

INTERNAL CONSTITUTION OF THE EARTH, edited by Beno Gutenberg. Prepared for National Research Council, this is a complete, thorough coverage of earth origins, continent formation, nature and behaviour of earth's core, petrology of crust, cooling forces in core, seismic and earthquake material, gravity, elastic constants, strain characteristics, similar topics. "One is filled with admiration . . . a high standard . . . there is no reader who will not learn something from this book," London, Edinburgh, Dublin, Philosophic Magazine. Largest Bibliography in print: 1127 classified items. Table of constants. 43 diagrams. 439pp. 6⅛ x 9¼. S414 Paperbound **$2.45**

THE BIRTH AND DEVELOPMENT OF THE GEOLOGICAL SCIENCES, F. D. Adams. Most thorough history of earth sciences ever written. Geological thought from earliest times to end of 19th century, covering over 300 early thinkers and systems; fossils and their explanation, vulcanists vs. neptunists, figured stones and paleontology, generation of stones, dozens of similar topics. 91 illustrations, including Medieval, Renaissance woodcuts, etc. 632 footnotes, mostly bibliographical. 511pp. 5⅜ x 8. T5 Paperbound **$2.00**

HYDROLOGY, edited by O. E. Meinzer, prepared for the National Research Council. Detailed, complete reference library on precipitation, evaporation, snow, snow surveying, glaciers, lakes, infiltration, soil moisture, ground water, runoff, drought, physical changes produced by water hydrology of limestone terranes, etc. Practical in application, especially valuable for engineers. 24 experts have created "the most up-to-date, most complete treatment of the subject," Am. Assoc. of Petroleum Geologists. 165 illustrations. xi + 712pp. 6⅛ x 9¼. S191 Paperbound **$2.95**

LANGUAGE AND TRAVEL AIDS FOR SCIENTISTS

SAY IT language phrase books

"SAY IT" in the foreign language of your choice! We have sold over ½ million copies of these popular, useful language books. They will not make you an expert linguist overnight, but they do cover most practical matters of everyday life abroad.

Over 1000 useful phrases, expressions, additional variants, substitutions.

Modern! Useful! Hundreds of phrases not available in other texts: "Nylon," "air conditioned," etc.

The ONLY inexpensive phrase book **completely indexed.** Everything is available at a flip of your finger, ready to use.

Prepared by native linguists, travel experts.

Based on years of travel experience abroad.

May be used by itself, or to supplement any other text or course. Provides a living element. Used by many colleges, institutions: Hunter College; Barnard College; Army Ordinance School, Aberdeen; etc.

Available, 1 book per language:

Danish (T818) 75¢
Dutch (T817) 75¢
English (for German-speaking people) (T801) 60¢
English (for Italian-speaking people) (T816) 60¢
English (for Spanish-speaking people) (T802) 60¢
Esperanto (T820) 75¢
French (T803) 60¢
German (T804) 60¢
Modern Greek (T813) 75¢
Hebrew (T805) 60¢

Italian (T806) 60¢
Japanese (T807) 75¢
Norwegian (T814) 75¢
Russian (T810) 75¢
Spanish (T811) 60¢
Turkish (T821) 75¢
Yiddish (T815) 75¢
Swedish (T812) 75¢
Polish (T808) 75¢
Portuguese (T809) 75¢

DOVER SCIENCE BOOKS

MONEY CONVERTER AND TIPPING GUIDE FOR EUROPEAN TRAVEL, C. Vomacka. Purse-size handbook crammed with information on currency regulations, tipping for every European country, including Israel, Turkey, Czechoslovakia, Rumania, Egypt, Russia, Poland. Telephone, postal rates; duty-free imports, passports, visas, health certificates; foreign clothing sizes; weather tables. What, when to tip. 5th year of publication. 128pp. 3½ x 5¼.   T260 Paperbound **60¢**

NEW RUSSIAN-ENGLISH AND ENGLISH-RUSSIAN DICTIONARY, M. A. O'Brien. Unusually comprehensive guide to reading, speaking, writing Russian, for both advanced, beginning students. Over 70,000 entries in new orthography, full information on accentuation, grammatical classifications. Shades of meaning, idiomatic uses, colloquialisms, tables of irregular verbs for both languages. Individual entries indicate stems, transitiveness, perfective, imperfective aspects, conjugation, sound changes, accent, etc. Includes pronunciation instruction. Used at Harvard, Yale, Cornell, etc. 738pp. 5⅜ x 8.   T208 Paperbound **$ 2.00**

PHRASE AND SENTENCE DICTIONARY OF SPOKEN RUSSIAN, English-Russian, Russian-English. Based on phrases, complete sentences, not isolated words—recognized as one of best methods of learning idiomatic speech. Over 11,500 entries, indexed by single words, over 32,000 English, Russian sentences, phrases, in immediately useable form. Shows accent changes in conjugation, declension; irregular forms listed both alphabetically, under main form of word. 15,000 word introduction covers Russian sounds, writing, grammar, syntax. 15 page appendix of geographical names, money, important signs, given names, foods, special Soviet terms, etc. Originally published as U.S. Gov't Manual TM 30-944. iv + 573pp. 5⅜ x 8.   T496 Paperbound **$2.75**

PHRASE AND SENTENCE DICTIONARY OF SPOKEN SPANISH, Spanish-English, English-Spanish. Compiled from spoken Spanish, based on phrases, complete sentences rather than isolated words—not an ordinary dictionary. Over 16,000 entries indexed under single words, both Castilian, Latin-American. Language in immediately useable form. 25 page introduction provides rapid survey of sounds, grammar, syntax, full consideration of irregular verbs. Especially apt in modern treatment of phrases, structure. 17 page glossary gives translations of geographical names, money values, numbers, national holidays, important street signs, useful expressions of high frequency, plus unique 7 page glossary of Spanish, Spanish-American foods. Originally published as U.S. Gov't Manual TM 30-900. iv + 513pp. 5⅝ x 8⅜.   T495 Paperbound **$1.75**

SAY IT CORRECTLY language record sets

The best inexpensive pronunciation aids on the market. Spoken by native linguists associated with major American universities, each record contains:

> 14 minutes of speech—12 minutes of normal, relatively slow speech, 2 minutes of normal conversational speed.

> 120 basic phrases, sentences, covering nearly every aspect of everyday life, travel—introducing yourself, travel in autos, buses, taxis, etc., walking, sightseeing, hotels, restaurants, money, shopping, etc.

> 32 page booklet containing everything on record plus English translations easy-to-follow phonetic guide.

> Clear, high-fidelity recordings.

> Unique bracketing systems, selection of basic sentences enabling you to expand use of SAY IT CORRECTLY records with a dictionary, to fit thousands of additional situations.

Use this record to supplement any course or text. All sounds in each language illustrated perfectly—imitate speaker in pause which follows each foreign phrase in slow section, and be amazed at increased ease, accuracy of pronounciation. Available, one language per record for

French	**Spanish**	**German**
Italian	**Dutch**	**Modern Greek**
Japanese	**Russian**	**Portuguese**
Polish	**Swedish**	**Hebrew**
English (for German-speaking people)		**English (for Spanish-speaking people)**

7" (33 1/3 rpm) record, album, booklet. **$1.00 each.**

SPEAK MY LANGUAGE: SPANISH FOR YOUNG BEGINNERS, M. Ahlman, Z. Gilbert. Records provide one of the best, most entertaining methods of introducing a foreign language to children. Within framework of train trip from Portugal to Spain, an English-speaking child is introduced to Spanish by native companion. (Adapted from successful radio program of N.Y. State Educational Department.) A dozen different categories of expressions,. including greeting, numbers, time, weather, food, clothes, family members, etc. Drill is combined with poetry and contextual use. Authentic background music. Accompanying book enables a reader to follow records, includes vocabulary of over 350 recorded expressions. Two 10" 33 1/3 records, total of 40 minutes. Book. 40 illustrations. 69pp. 5¼ x 10½.   T890 The set **$4.95**

LISTEN & LEARN language record sets

LISTEN & LEARN is the only extensive language record course designed especially to meet your travel and everyday needs. Separate sets for each language, each containing three 33 1/3 rpm long-playing records—1 1/2 hours of recorded speech by eminent native speakers who are professors at Columbia, New York U., Queens College.

Check the following features found only in LISTEN & LEARN:

Dual language recording. 812 selected phrases, sentences, over 3200 words, spoken first in English, then foreign equivalent. Pause after each foreign phrase allows time to repeat expression.

128-page manual (196 page for Russian)—everything on records, plus simple transcription. Indexed for convenience. Only set on the market completely indexed.

Practical. No time wasted on material you can find in any grammar. No dead words. Covers central core material with phrase approach. Ideal for person with limited time. Living, modern expressions, not found in other courses. Hygienic products, modern equipment, shopping, "air-conditioned," etc. Everything is immediately useable.

High-fidelity recording, equal in clarity to any costing up to $6 per record.

"Excellent . . . impress me as being among the very best on the market," Prof. Mario Pei, Dept. of Romance Languages, Columbia U. "Inexpensive and well done . . . ideal present," Chicago Sunday Tribune. "More genuinely helpful than anything of its kind," Sidney Clark, well-known author of "All the Best" travel books.

UNCONDITIONAL GUARANTEE. Try LISTEN & LEARN, then return it within 10 days for full refund, if you are not satisfied. It is guaranteed after you actually use it.

6 modern languages—FRENCH, SPANISH, GERMAN, ITALIAN, RUSSIAN, or JAPANESE *—one language to each set of 3 records (33 1/3 rpm). 128 page manual. Album.

Spanish	the set $4.95	**German**	the set $4.95	**Japanese***	the set $5.95
French	the set $4.95	**Italian**	the set $4.95	**Russian**	the set $5.95

* Available Oct. 1959.

TRÜBNER COLLOQUIAL SERIES

These unusual books are members of the famous Trübner series of colloquial manuals. They have been written to provide adults with a sound colloquial knowledge of a foreign language, and are suited for either class use or self-study. Each book is a complete course in itself, with progressive, easy to follow lessons. Phonetics, grammar, and syntax are covered, while hundreds of phrases and idioms, reading texts, exercises, and vocabulary are included. These books are unusual in being neither skimpy nor overdetailed in grammatical matters, and in presenting up-to-date, colloquial, and practical phrase material. Bilingual presentation is stressed, to make thorough self-study easier for the reader.

COLLOQUIAL HINDUSTANI, A. H. Harley, formerly Nizam's Reader in Urdu, U. of London. 30 pages on phonetics and scripts (devanagari & Arabic-Persian) are followed by 29 lessons, including material on English and Arabic-Persian influences. Key to all exercises. Vocabulary. 5 x 7½. 147pp. Clothbound **$1.75**

COLLOQUIAL ARABIC, DeLacy O'Leary. Foremost Islamic scholar covers language of Egypt, Syria, Palestine, & Northern Arabia. Extremely clear coverage of complex Arabic verbs & noun plurals; also cultural aspects of language. Vocabulary. xviii + 192pp. 5 x 7½. Clothbound **$1.75**

COLLOQUIAL GERMAN, P. F. Doring. Intensive thorough coverage of grammar in easily-followed form. Excellent for brush-up, with hundreds of colloquial phrases. 34 pages of bilingual texts. 224pp. 5 x 7½. Clothbound **$1.75**

COLLOQUIAL SPANISH, W. R. Patterson. Castilian grammar and colloquial language, loaded with bilingual phrases and colloquialisms. Excellent for review or self-study. 164pp. 5 x 7½. Clothbound **$1.75**

COLLOQUIAL FRENCH, W. R. Patterson. 16th revised edition of this extremely popular manual. Grammar explained with model clarity, and hundreds of useful expressions and phrases; exercises, reading texts, etc. Appendixes of new and useful words and phrases. 223pp. 5 x 7½. Clothbound **$1.75**

DOVER SCIENCE BOOKS

COLLOQUIAL PERSIAN, L. P. Elwell-Sutton. Best introduction to modern Persian, with 90 page grammatical section followed by conversations, 35 page vocabulary. 139pp.   Clothbound **$1.75**

COLLOQUIAL CZECH, J. Schwarz, former headmaster of Lingua Institute, Prague. Full easily followed coverage of grammar, hundreds of immediately useable phrases, texts. Perhaps the best Czech grammar in print. "An absolutely successful textbook," JOURNAL OF CZECHO-SLOVAK FORCES IN GREAT BRITAIN. 252pp. 5 x 7½.   Clothbound **$2.50**

COLLOQUIAL RUMANIAN, G. Nandris, Professor of University of London. Extremely thorough coverage of phonetics, grammar, syntax; also included 70 page reader, and 70 page vocabulary. Probably the best grammar for this increasingly important language. 340pp. 5 x 7½.
Clothbound **$2.50**

COLLOQUIAL ITALIAN, A. L. Hayward. Excellent self-study course in grammar, vocabulary, idioms, and reading. Easy progressive lessons will give a good working knowledge of Italian in the shortest possible time. 5 x 7½.   Clothbound **$1.75**

MISCELLANEOUS

TREASURY OF THE WORLD'S COINS, Fred Reinfeld. Finest general introduction to numismatics; non-technical, thorough, always fascinating. Coins of Greece, Rome, modern countries of every continent, primitive societies, such oddities as 200-lb stone money of Yap, nail coinage of New England; all mirror man's economy, customs, religion, politics, philosophy, art. Entertaining, absorbing study; novel view of history. Over 750 illustrations. Table of value of coins illustrated. List of U.S. coin clubs. 224pp. 6½ x 9¼.
T433 Paperbound **$1.75**

ILLUSIONS AND DELUSIONS OF THE SUPERNATURAL AND THE OCCULT, D. H. Rawcliffe. Rationally examines hundreds of persistent delusions including witchcraft, trances, mental healing, peyotl, poltergeists, stigmata, lycanthropy, live burial, auras, Indian rope trick, spiritualism, dowsing, telepathy, ghosts, ESP, etc. Explains, exposes mental, physical deceptions involved, making this not only an exposé of supernatural phenomena, but a valuable exposition of characteristic types of abnormal psychology. Originally "The Psychology of the Occult." Introduction by Julian Huxley. 14 illustrations. 551pp. 5⅜ x 8.
T503 Paperbound **$2.00**

HOAXES, C. D. MacDougall. Shows how art, science, history, journalism can be perverted for private purposes. Hours of delightful entertainment, a work of scholarly value, often shocking. Examines nonsense news, Cardiff giant, Shakespeare forgeries, Loch Ness monster, biblical frauds, political schemes, literary hoaxers like Chatterton, Ossian, disumbrationist school of painting, lady in black at Valentino's tomb, over 250 others. Will probably reveal truth about few things you've believed, will help you spot more easily the editorial "gander" or planted publicity release. "A stupendous collection . . . and shrewd analysis," New Yorker. New revised edition. 54 photographs. 320pp. 5⅜ x 8.   T465 Paperbound **$1.75**

YOGA: A SCIENTIFIC EVALUATION, Kovoor T. Behanan. Book that for first time gave Western readers a sane, scientific explanation, analysis of yoga. Author draws on laboratory experiments, personal records of year as disciple of yoga, to investigate yoga psychology, physiology, "supernatural" phenomena, ability to plumb deepest human powers. In this study under auspices of Yale University Institute of Human Relations, strictest principles of physiological, psychological inquiry are followed. Foreword by W. A. Miles, Yale University. 17 photographs. xx + 270pp. 5⅜ x 8.   T505 Paperbound **$1.65**